Say It with Shapes and Numbers

Games, projects, and activities that mix in math for ages 3–6

Marlene Kliman, Valerie Martin, Nuria Jaumot-Pascual

TERC
2067 Massachusetts Ave
Cambridge, MA 02140
http://www.terc.edu

SAY IT WITH SHAPES AND NUMBERS
Author: Marlene Kliman
Design, illustration, and composition: Valerie Martin
Developmental editor: Nuria Jaumot-Pascual

Tumblehome Learning
P.O. Box 171386
Boston, MA 02117
www.tumblehomelearning.com

This book was developed at TERC and is based on work funded in part and supported by The National Science Foundation, under grants DRL-07145537, ESI-0406675, and ESI-9901289. Any opinions, findings, and conclusions or recommendations expressed in this material are those of the authors and do not necessarily reflect the views of The National Science Foundation.

TERC is a not-for-profit education research and development organization dedicated to improving mathematics, science, and technology teaching and learning.

Printed in Taiwan
10 9 8 7 6 5 4 3 2 1

Library of Congress Control Number: 2014945609

ISBN 978-0-9897924-6-2

© 2014 TERC

All rights reserved. No part of this book may be used or reproduced, stored in a retrieval system, or transmitted by any means—electronic, mechanical, photocopying, recording, or in any manner whatsoever—without written permission from the authors except in the case of brief quotations in critical articles and reviews.

Table of Contents

Introduction v

Making the most of SAY IT WITH SHAPES AND NUMBERS vii

Games

Match 2
Share the Teddy Bears 4
Same or Different? 5
Secret Card 6
Piggy Bank 8
Piggy Bank Boards 9
Dinosaur Dash 10
Dinosaur Dash Boards 11
Hide, Share, and Compare 13
Empty the Toy Box 14
Empty the Toy Box Board 15
Flip and Match 16
Ladybug Card Deck 17

Projects and Crafts

Revise Your Size 34
Strawberries and Blueberries 35
Soaring Towers 36
Picture Book 37
Fill It Up 38
Cityscape 39
Watch Me Grow 40
Make a Mask 41
Chain Challenges 42
Towering Toothpicks 43
Make It Morse 44
Morse Code Chart 45
Hands on the Wall 46
Copy Cat 47
Say It with Shapes 48
Picture Poems 50
Word Shapes 51
Blank Shapes 57

Food and Water

Fill It or Spill It 72
Snack Station 73
Water Wonders 74
One for You, One for Me 75
Pretend Picnic 76
What's Inside? 77
In the Bag 78
My Height in Boxes 79
The Counting Chef 80

Good for Groups

Catch the Beat 82
Group Up 83
Stand and Vote 84
Quick Questions 85
Toy Store 86
Who's Here? 88
Line Up 89
Treasure Hunt 90

Anytime, Anywhere

Words on the Wall 92
Yes, No, Maybe 93
How Far Can You Go? 94
What's on the Page? 95
Seeing Shapes 96
Take Two 98

All Year Round

Seasons 99

Math Connections

Introduction 105
NAEYC/NCTM and Common Core Chart 106

Additional Letters and Shapes 111

© 2014 TERC • Cambridge, MA

Say It with Shapes and Numbers

Games, projects, and activities that mix in math for ages 3–6

Play games, build towers, move to the beat, and go on a pretend picnic—all with math!

This book contains over 300 ways to mix in math.

Games

Play to match colors and numbers, to compare, sort, and count, and to add and subtract. Includes Game boards and Ladybug Card Deck. Durable Ladybug Card Deck with additional games are also available at www.tumblehomelearning.com.

Projects and Crafts

Build, design, and create with projects and crafts that use everyday materials. Includes letters and English and Spanish words in colorful geometric shapes to use for projects that combine geometry, patterns, and literacy.

Food and Water

Ideas to investigate and games to play in the kitchen, at snack time, or around water.

Good for Groups

Group games, party favorites, icebreakers, and circle time activities for indoors and out.

Anytime, Anywhere

Activities to do and games to play wherever you are: in the car, on the bus, in a waiting room, or at the dinner table.

All Year Round

Ideas for mixing math into seasons, holidays, and special events all year round.

Aligned with the NAEYC/NCTM Joint Position Statement on Early Childhood Mathematics and the Kindergarten Common Core Math Standards. Developed by the MIXING IN MATH group at TERC, an education non-profit, and based on research funded in part by the National Science Foundation.

Why did we write SAY IT WITH SHAPES AND NUMBERS?

We believe that creativity, play, and socializing are important ingredients in learning just about anything. This book is designed to put those ingredients into learning math.

We started with the activities, games, and projects that young children enjoy in childcare programs, at preschool, at public libraries, and at home. We highlighted the inherent math with things to talk about, and sometimes we added a mathematical twist. To ensure that our materials were engaging and enriching, we piloted them in a wide range of settings. Independent research showed that children and adults gained math skills, confidence in their math abilities, understanding of the role of math in everyday life, and positive attitudes toward math. To find out more about this research, see http://mixinginmath.terc.edu/aboutMiM/index.php.

SAY IT WITH SHAPES AND NUMBERS is based on nearly 15 years of development and research funded in part by The National Science Foundation.

Who is this book for?

Everyone! It's for young children and their families, for mathophobes and mathophiles, and for parents, childcare providers, librarians, and teachers. The games, projects, and activities are geared toward children ages 3-6, but older children will also enjoy and find challenge in them. Some are perfect for one or two children with an adult or older child, others work well with a group at home, at a party or family event, at a childcare program, at school, or just about anywhere.

What math is in the book?

The games, projects, and activities in this book span the key topics addressed in the NAEYC/NCTM Joint Position Statement on Early Childhood Mathematics and the Kindergarten Common Core State Standards for Mathematics. See pp. 105-109 for more detail. Many of the ideas in this book are interdisciplinary, including topics in literacy, arts, science, social studies, and engineering.

Thank you!

The authors of the book, Marlene Kliman, Valerie Martin, and Nuria Jaumot-Pascual, are very grateful to Martha Merson of TERC for her contributions including ideas behind our word shape projects, to Lily Ko of TERC for her input and support, to our external evaluators for evidence-based insights (Char Associates, Miller-Midzik Research Associates, and Program Evaluation Research Group), to Laura DeSantis for amazing images, and to the many childcare and after-school providers, librarians, parents, and children who have collaborated with us over the years. We extend our appreciation to TERC for providing a home for our MIXING IN MATH projects. Marlene would like to thank her daughters Clara and Chloe for helping her mix in math from the start and for providing a reality check on math at home.

Making the most of SAY IT WITH SHAPES AND NUMBERS

This book contains hundreds of games, projects, and activities. You can do them in any order. Look for the information below in each game, project, and activity.

Level. Each is marked with one or more of Easy, Medium, and Hard. Levels reflect "typical" 3-4 year old (Easy), 4-5 year old (Medium), and 5-6 year old (Hard) cognitive, social, language, fine motor, and other skills. Keep in mind that young children vary widely in their abilities. Some people start with Easy for almost any age and then move up as needed.

Note: Levels that only appear in Variations are in parentheses.

Group size. Some projects and activities can be done alone; others are best done with a group. Games require more than one player.

Materials. Some involve no materials; others rely on common materials.

Variations. Includes ways to vary and adapt for different levels of challenge.

Talk About. Offers ideas on what to talk over in order to support math skills.

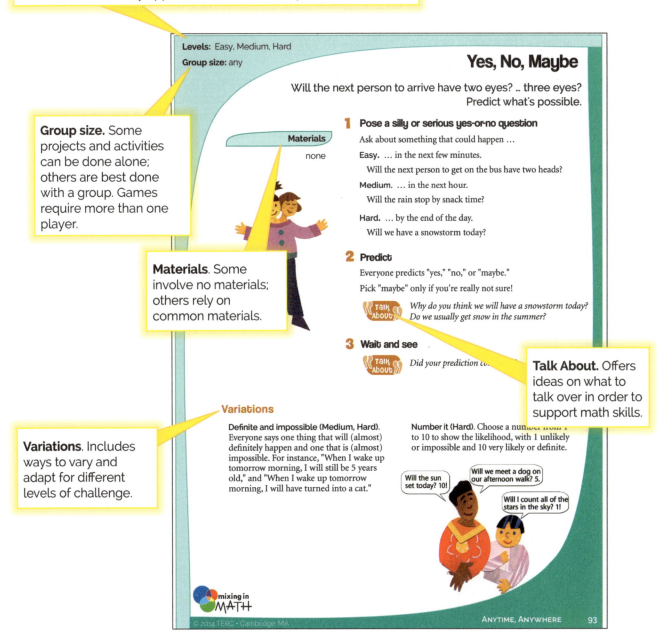

© 2014 TERC • Cambridge, MA — INTRODUCTION — vii

About the Authors

Marlene Kliman, Senior Scientist and Director of the Mixing in Math group at TERC, brings 30 years experience developing research-based resources for children's math learning in and out of school. A Principal Investigator of out-of-school math projects funded by The National Science Foundation, she has collaborated with a wide range of educational organizations including after-school programs, public libraries, and family literacy centers. She formerly taught math to pre-service elementary grades teachers at Lesley University. Marlene completed her undergraduate studies in mathematics at Harvard and her graduate studies in learning and epistemology at MIT.

Valerie Martin, Senior Web and Graphic Designer at TERC, specializes in conveying math and science concepts in a clear and visually appealing manner. She designs web- and print-based curricula, games, and educational resources for a wide range of audiences, including children, parents, preschool teachers, adult education teachers, and museum educators. Valerie holds degrees from SUNY Binghamton in French and German literature, with further studies in graphic design and website design and development.

Nuria Jaumot-Pascual, Senior Research Associate at TERC, has 20 years experience as a bilingual (Spanish/English) preschool and after-school teacher, staff developer, and educational researcher in Spain, Central America, and the United States. Nuria holds degrees in out of school education (Universitat de Barcelona), anthropology (University of Texas, Austin), educational psychology (Universitat Oberta de Catalunya), and organizational behavior (Harvard). She is currently a doctoral student in education at the University of Georgia.

Games

Play to match colors and numbers, to compare, sort, and count, and to add and subtract.

Contents

Match 2
Share the Teddy Bears 4
Same or Different? 5
Secret Card 6
Piggy Bank 8
Piggy Bank Boards 9
Dinosaur Dash 10
Dinosaur Dash Boards 11
Hide, Share, and Compare 13
Empty the Toy Box 14
Empty the Toy Box Board 15
Flip and Match 16
Ladybug Card Deck 17

Games in other sections

Projects and Crafts
Copy Cat 47
Food and Water
In the Bag 78
Good for Groups
Treasure Hunt 92
Anytime, Anywhere
Words on the Wall 92
Seeing Shapes 96

 Durable Ladybug Card Deck and additional games are available at www.tumblehomelearning.com.

© 2014 TERC • Cambridge, MA

Levels: Easy (Medium, Hard)

Match

Match the number of the top card in the pile.

Set up

Mix up the cards. Pile them face down to make the Draw pile.

Turn over the top card in the Draw pile. Place it face up to start the Match pile.

 How many dots on the card in the Match pile?

Play

Take turns. On your turn:

1 Turn over the top card in the Draw pile

 How many dots are on the card you turned over?

2 Look for a match

Does your card have the same number of dots as the card on top of the Match pile?

Yes: Place your card face up on the Match pile.

No: Discard your card.

Your turn is over.

End the game

Keep taking turns until the Draw pile is empty. The game is over.

Find the winner (optional)

Keep track of how many matches you make. The player who made the most matches wins.

Materials

Per game

Ladybug Card Deck, cards 0–4

 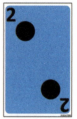

Draw pile   Match pile

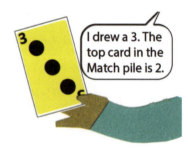

I drew a 3. The top card in the Match pile is 2.

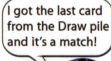

I got the last card from the Draw pile and it's a match!

Players: 2

Match, cont'd

Variations

Color match (Easy). Match color instead of number. If the top card in the Match pile is green, you can match with any green card.

Play to 6 (Medium) or Play wild (Hard). Use cards 0-6. For even more challenge, include Wild Cards.

A Wild Card can be any of the four card colors and any number from 0 to 6.

Match number or color (Hard). On each turn, match the color or the number of the top card in the Match pile. If the top card is a blue 2, you can match with any blue card or any 2.

I don't have a 2 but I have another blue card so I can use that.

Match number or color in a row (Hard). Instead of making a Match pile, make a Match Row. Start with one card face up. On each turn, look for a match on the left or right side of the row. Use cards 0-6 and Wild Cards.

I can put my pink 3 on either end.

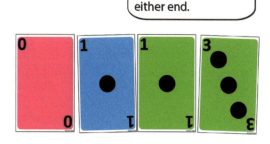

Play in two directions (Hard). Match across or up and down. Start with one card face up. On each turn, try to match the number of dots or the color. You can place your card one of these ways:
- above the match
- below the match
- next to the match.

Use cards 0-6 and Wild Cards.

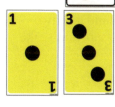

GAMES 3

Levels: Easy (Medium, Hard)
Players: 2

Share the Teddy Bears

Draw a card to find out how many teddy bears to take. If you take two, share the wealth!

Set up

Mix up the cards. Place the cards face down to make the Draw pile.

Play

Take turns. On your turn:

1 Turn over the top card in the Draw pile

2 Get your teddy bears

If you draw a 1, take one teddy bear.

If you draw a 2, take two teddy bears and:
- keep one teddy bear
- give one to the other player

 How many teddy bears do you have in all?

Your turn is over.

End the game

Keep taking turns until the Draw pile is empty. The game is over.

Find the winner (optional)

The player with the most teddy bears wins.

Materials

Per game

Ladybug Card Deck, cards 1-2 in three colors (total of six cards)

9 teddy bear or other counters

I drew a 2 card so I keep 1 teddy bear and give 1 to Sam.

Variations

Share the teddy bears to 4 (Medium). Play with cards 0, 1, 2, and 4 in two colors (total of eight cards), and 14 teddy bears. If you draw a 4, take two teddy bears and give two to the other player. If you draw a 0, don't take any teddy bears.

Share the teddy bears to 6 (Hard). Play with cards 0, 1, 2, 4, and 6 in two colors (total of ten cards), and 26 teddy bears. If you draw a 4, take two teddy bears and give two to the other player. If you draw a 6, take three and give three to the other player. If you draw 0, don't take any teddy bears.

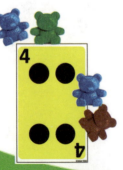

Levels: Easy, Medium (Hard)
Players: 2

Same or Different?

Compare handfuls of beads to see if both players grabbed the same amount.

Set up

Each player chooses a bowl of objects to use.

Materials

Per game

one bowl of blue beads or counters and one bowl of red beads or counters

Easy. Use objects sized so a child can hold 2-4 in one hand.

Medium. Use objects sized so a child can hold 5-6 in one hand.

Play

1 Grab a handful

Each player grabs a handful from a different bowl.

2 Match them up

Lay out your objects and pair them up.

 Can you match each red one with a blue one? Are there any left over?

3 Same or different?

Decide if your handfuls have the same or a different amount.

End the game

Put your objects back in the bowl. The game is over.

Find the winner (optional)

At the start of the game, decide who will be Same and who will be Different. If the handfuls have the same amount, Same wins. Otherwise, Different wins.

Variations

Match up (Easy, Medium). Try this with plastic bottle caps in one bowl and beads in the other. Match them up by placing the beads in the bottle caps.

Count it out (Easy, Medium). Count to find out how many each person has.

Save the game (Easy, Medium). Glue your objects on paper to show how you matched them up.

Total of 10 (Hard). Play until you grab handsful that total exactly 10.

Levels: Medium, Hard

Secret Card

Ask yes-or-no questions to identify the Secret Card.

Set up

Spread out the cards face up.

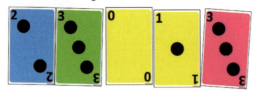

Decide who will pick the first Secret Card. Each player will have a turn.

Materials

Per game

Medium. Ladybug Card Deck, any 3-4 cards.

Hard. Ladybug Card Deck, any 5-6 cards.

Play

1 One player secretly picks a card

Do not remove the card from the layout.

2 The other player asks a yes-or-no question

Ask a question to try to identify the secret card. Ask about features of the cards. Do not ask if a certain card is the secret one.

   Is it blue?
Does it have two dots?
Does it have more than two dots?

3 Answer and rule out cards

The player who picked the secret card answers the question and removes any cards that were ruled out. For instance, if someone asks "Is it blue?" and the answer is no, remove the blue cards.

   How do you know which cards to remove?

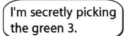

Players: 2 (see Variations for more)

Secret Card, cont'd

4 Keep asking and answering

Play until the secret card is the only card left.

End the game

The game is over when both players have had a turn to pick the secret card.

Find the winner (optional)

Keep track of how many questions you ask to identify the secret card. The player who identifies the card with the fewest questions wins.

Variations

Tell someone (Medium, Hard). After you secretly pick a card, whisper your choice to an adult. The adult can remind you if you forget what you picked.

Secret button or Secret shape (Medium, Hard). Play with buttons or with word, letter, or blank shapes on pp. 51-70.

Ask yes-or-no questions about colors, number of holes, size, or letters.

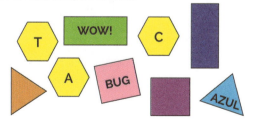

Play with a group (Hard). Play with three or four players. One picks the secret card; the others take turns asking yes-or-no questions. The player who narrows the layout down to the secret card wins.

Secret person (Medium, Hard). Play with 4-10 people. An adult secretly picks one person in the group. To narrow down to the secret person, players ask yes-or-no questions.

GAMES 7

Levels: Easy (Medium, Hard)
Players: 2

Piggy Bank

Take turns putting pennies in the piggy bank. Can you fill the bank?

Set up

Mix up the cards. Pile them face down next to the piggy bank to start the Draw pile.

Play

Take turns. On your turn:

1 Turn over the top card in the Draw pile

2 Get your pennies

If you draw 1, take one penny.

If you draw 2, take two pennies.

 How many dots on the card? How many pennies do you take?

3 Put your pennies in the piggy bank

Put one penny in each space in the piggy bank. If you have an extra penny, put it aside. Your turn is over.

 How many spaces are full? Are any empty?

End the game

Keep taking turns until the piggy bank is full. The game is over.

Find the winner (optional)

The player to put the last penny in the piggy bank wins.

Materials

Per game

Ladybug Card Deck, cards 1-2

5 pennies

Piggy Bank 4 Board (p. 9)

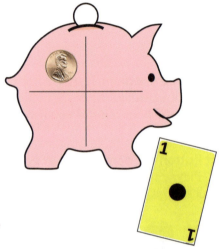

Variations

Piggy bank 6 (Medium). Play with cards 0-3. Use the Piggy Bank 6 Board (p. 9).

Piggy bank 10 (Hard). Play with cards 0-6 and Wild Cards. Draw a piggy bank board with 10 spaces.

Nickels and dimes (Hard). Play with nickels or dimes. Count by 5s or 10s to find how much you have in all.

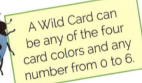
A Wild Card can be any of the four card colors and any number from 0 to 6.

8 GAMES © 2014 TERC • Cambridge, MA

Piggy Bank 4 Board

Piggy Bank 6 Board

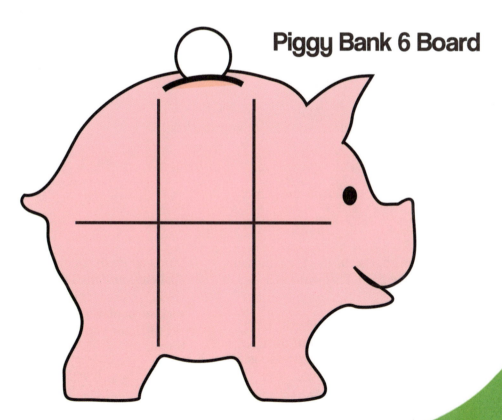

GAMES

Levels: Medium (Hard)
Players: 2

Dinosaur Dash

Take turns helping the dinosaur dash to 10.

Set up

Mix up the cards. Place the cards face down to make the Draw pile. Put the dinosaur on number 1 on the game board.

Materials

Per game
- Ladybug Card Deck, cards 1-3
- Dinosaur Dash 10 Board (p. 11)
- dinosaur or other counter

Play

Take turns. On your turn:

1 Pick a card

2 Dash

 What number does your card show?

Read the number on your card. Help the dinosaur dash that many spaces.

If the dinosaur:
- reaches 10 (or more), the game is over.
- reaches a number less than 10, your turn is over.

 How far is the dinosaur from 10?

2. I'll make the dino dash from 3 to 5.

End the game

Keep taking turns. The game is over when the dinosaur reaches 10.

Find the winner (optional)

The player to get the dinosaur to 10 wins the game.

Variations

Dinosaur dash 20 (Hard). Use the Dinosaur Dash 20 Board (p. 11). Play with cards 0-3 or cards 0-6.

Double dinosaurs (Medium, Hard). Each player has a dinosaur. Move your own dinosaur on your turn. First person to get to the end of the board wins.

Dinosaur dash 100 (Hard). For extra challenge, use the Dinosaur Dash 100 Board (p. 12). Use cards 0-6 and Wild Cards. On each turn, players draw two cards. They find the total and move that many spaces on the board.

A Wild Card can be any of the four card colors and any number from 0 to 6.

Dinosaur Dash 10 Board

| 1 | 2 | 3 | 4 | 5 | 6 | 7 | 8 | 9 | 10 |

Dinosaur Dash 20 Board

| 1 | 2 | 3 | 4 | 5 | 6 | 7 | 8 | 9 | 10 |
| 11 | 12 | 13 | 14 | 15 | 16 | 17 | 18 | 19 | 20 |

© 2014 TERC • Cambridge, MA

GAMES 11

Dinosaur Dash 100 Board

1	2	3	4	5	6	7	8	9	10
11	12	13	14	15	16	17	18	19	20
21	22	23	24	25	26	27	28	29	20
31	32	33	34	35	36	37	38	39	40
41	42	43	44	45	46	47	48	49	50
51	52	53	54	55	56	57	58	59	60
61	62	63	64	65	66	67	68	69	70
71	72	73	74	75	76	77	78	79	80
81	82	83	84	85	86	87	88	89	90
91	92	93	94	95	96	97	98	99	100

Levels: Easy (Medium, Hard)
Players: 2

Hide, Share, and Compare

One, two, three—share and compare. Do your fingers match your partner's?

Materials

none

Set up
Sit or stand facing your partner.

Play

1 Each player secretly picks a number

Secretly pick 1, 2, or 3. Hide your secret number of fingers behind your back so your partner can't see.

2 One, two, three, show!

On the count of three, hold out your fingers.

3 Same or different?

 Do we have the same number of fingers out?

Compare to see if you match.

End the game

Shake out your fingers! The game is over.

Find the winner (optional)

At the start of the game, decide who will be Same and who will be Different. If you match, Same wins. Otherwise, Different wins. Then, switch roles and play again.

Variations

Play with counters (Easy, Medium, Hard). Instead of using your fingers, hold, share, and compare a secret number of counters.

Play your hand (Medium). Play with 1 to 5 fingers.

Play two hands (Hard). Play with 0 to 10 fingers.

Closest match (Hard). Mix up Ladybug cards 0-6 and pile them face down. Each player holds out between 0 and 6 fingers. Then, they turn over the top card in the deck. The player closest to number on the card wins.

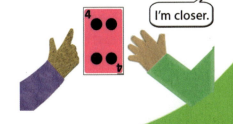

© 2014 TERC • Cambridge, MA

GAMES 13

Empty the Toy Box

Teddy bears are escaping the toy box. Try to catch the last one.

Set up

Put one teddy bear in each space in the toy box.

Materials

Per game

7 teddy bear or other counters

Empty the Toy Box Board (p. 15)

Play

Take turns. On your turn:

1 Plan

On each turn, players will take one or two teddy bears. Try to plan ahead so you take the last one.

 If you take two, how many will be left? Could you get the last one?

2 Take one or two teddy bears

 How many teddy bears are left in the toy box?

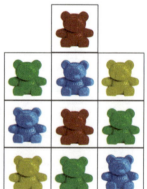

There are 4 left. If I take 2, Mellie will take the last 2 and the game will be over.

End the game

Keep taking turns until the toy box is empty. The game is over.

Find the winner (optional)

The player who takes the last teddy bear wins.

Variations

Fill the toy box (Medium, Hard). Start with seven teddy bears and an empty toy box. On each turn, players can put one or two teddy bears in the toy box. The player to put in the last teddy bear wins.

Ten teddy bears (Hard). Make a "toy box" with ten squares and put a teddy bear in each one. On each turn, take one, two, or three teddy bears. The player who takes the last teddy bear wins.

14   GAMES   © 2014 TERC • Cambridge, MA

Empty the Toy Box Board

Levels: Easy, Medium, Hard
Players: 2

Flip and Match

Flip two cards to make a match with this memory game.

Set up

Mix the cards up and lay them out face down:

Easy. Two rows of three cards each.
Medium. Two rows of five cards each.
Hard. Four rows of four cards each.

Materials

Per game

Easy. Ladybug Card Deck, cards 1-3, in two colors (total of 6 cards).

Medium. Ladybug Card Deck, cards 0-4, in two colors (total of 10 cards).

Hard. Ladybug Card Deck, cards 0-3, in four colors (total of 16 cards).

Play

Take turns. On your turn:

1 Turn over two cards

 How many dots on the card you turned over? … Do you remember where the other 3 card is?

2 Look for a match

Do both cards have the same number of dots?

Yes: Take both cards out of the layout. Keep the pair.

No: Turn the cards face down in the layout. Make sure to put them back where you found them.

Your turn is over.

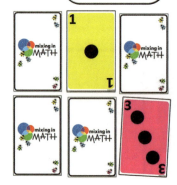

"Not the same number of dots. No match."

End the game

Keep taking turns. The game is over when no cards are left in the layout.

Find the winner (optional)

The player who made the most pairs wins.

Variations

Match colors (Easy). Make pairs with matching colors instead of matching numbers.

Exact match (Easy, Medium). Use two Ladybug Card Decks. Include duplicate cards in the layout. Match color and number to make a pair.

Find five (Hard). Lay out cards 0-5 in two colors. Look for pairs of cards that add up to 5. For even more challenge, use Wild Cards.

A Wild Card can be any of the four card colors and any number from 0 to 6.

16   GAMES   © 2014 TERC • Cambridge, MA

Ladybug Card Deck

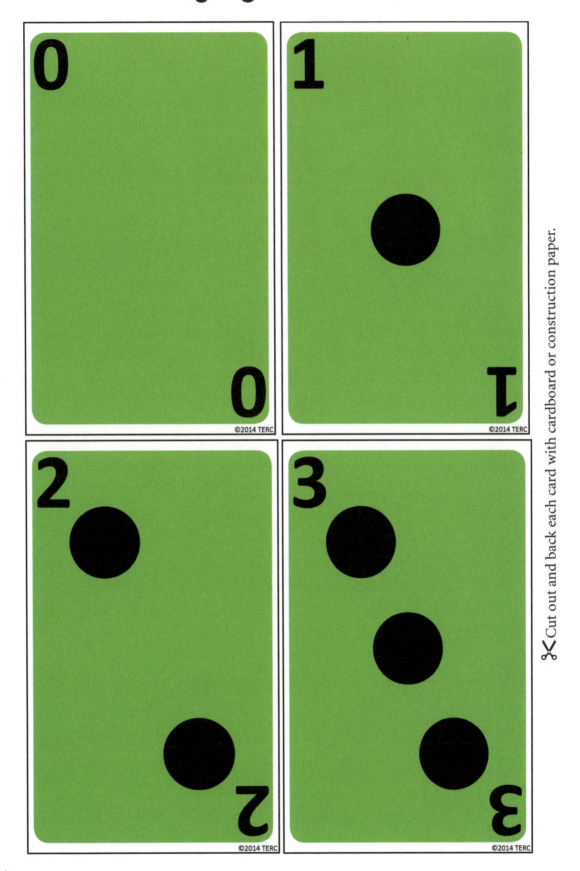

⭐ Ladybug Card Deck is available for purchase as a durable deck with additional games at www.tumblehomelearning.com.

Ladybug Card Deck

✂ Cut out and back each card with cardboard or construction paper.

⭐ Ladybug Card Deck is available for purchase as a durable deck with additional games at www.tumblehomelearning.com.

Ladybug Card Deck

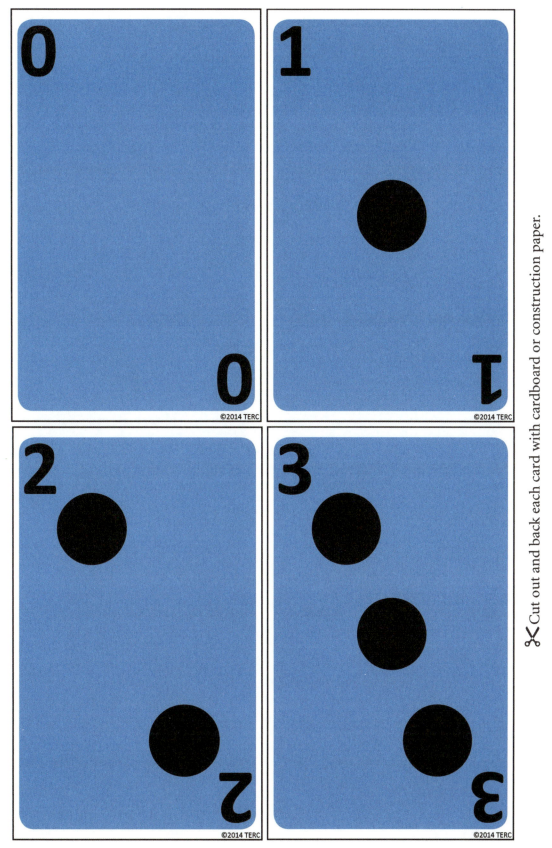

Cut out and back each card with cardboard or construction paper.

⭐ Ladybug Card Deck is available for purchase as a durable deck with additional games at www.tumblehomelearning.com.

Ladybug Card Deck

Cut out and back each card with cardboard or construction paper.

⭐ Ladybug Card Deck is available for purchase as a durable deck with additional games at www.tumblehomelearning.com.

© 2014 TERC • Cambridge, MA   GAMES   23

Ladybug Card Deck

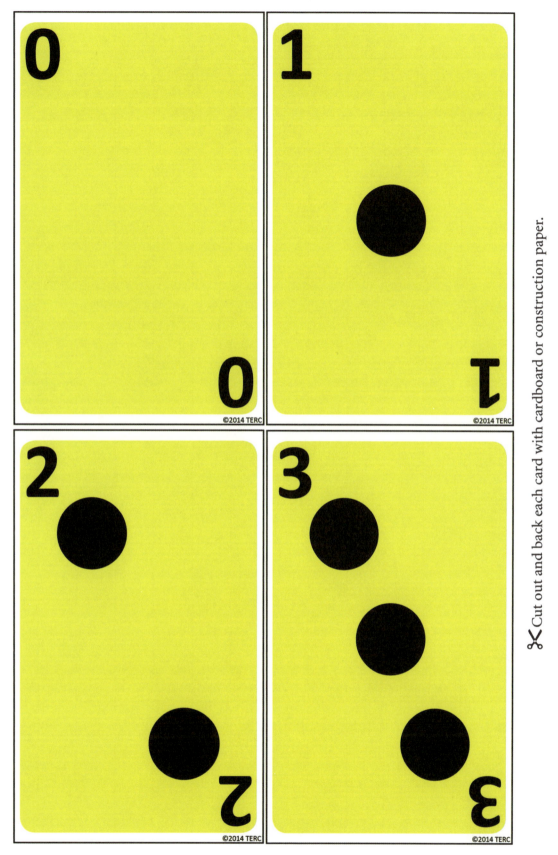

⭐ Ladybug Card Deck is available for purchase as a durable deck with additional games at www.tumblehomelearning.com.

Ladybug Card Deck

⭐ Ladybug Card Deck is available for purchase as a durable deck with additional games at www.tumblehomelearning.com.

Ladybug Card Deck

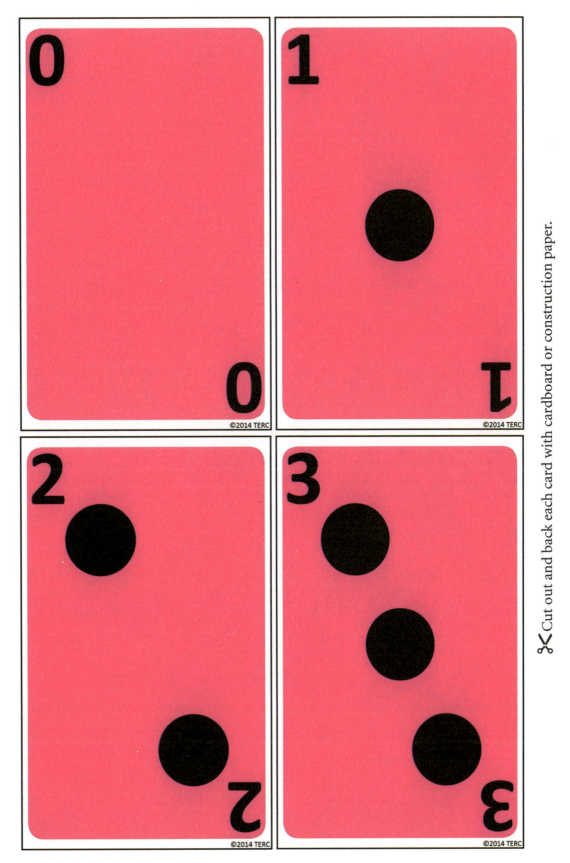

✂ Cut out and back each card with cardboard or construction paper.

⭐ Ladybug Card Deck is available for purchase as a durable deck with additional games at www.tumblehomelearning.com.

GAMES 29

Ladybug Card Deck

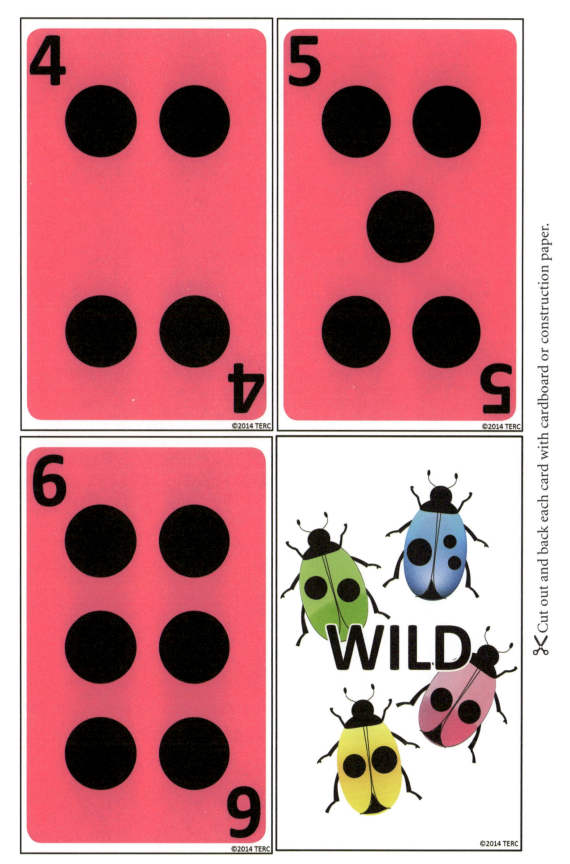

✂ Cut out and back each card with cardboard or construction paper.

⭐ Ladybug Card Deck is available for purchase as a durable deck with additional games at www.tumblehomelearning.com.

Projects and Crafts

Build, design, and create with projects and crafts that use everyday materials.

Contents

Revise Your Size	34
Strawberries and Blueberries	35
Soaring Towers	36
Picture Book	37
Fill It Up	38
Cityscape	39
Watch Me Grow	40
Make a Mask	41
Chain Challenges	42
Towering Toothpicks	43
Make It Morse	44
Morse Code Chart	45
Hands on the Wall	46
Copy Cat	47
Say It with Shapes	48
Picture Poems	50
Word Shapes	51
Letter Shapes	59
Blank Shapes	61
Additional Shapes and Letters	120

Projects and Crafts in other sections

Food and Water

My Height in Boxes	79
The Counting Chef	80

Good for Groups

Catch the Beat (Variation, Draw the beat)	45
Treasure Hunt (Variation, Map it)	47

© 2014 TERC • Cambridge, MA

Revise Your Size

Levels: Easy, Medium, Hard
Group size: 1 per height strip

Make a paper strip as tall as you are. Repeat in a few months. Are you getting taller?

1 Make a height strip

An adult cuts a strip of paper as tall as you are.

2 Decorate

Personalize your height strip.

Each strip needs a name and a date.

3 Revise your size

Save your height strip. Make another one in a few months. Tack the strips on the wall, so the bottoms are lined up with the floor. Then compare:

Easy. *Which strip goes up to the top of your head? Which one just goes up to your eyes?*

Medium. *Do you think you grew? Why or why not?*

Hard. *Why do we need to line up the bottom of the strips to compare them?*

Materials

Per height strip

adding machine tape or wrapping paper cut into a strip a few inches wide and at least 4 feet long

To share

art supplies or stickers for decorating strips

For an adult

scissors

Variations

Height museum (Hard). Try this with a group of 4-6. Line up your height strips in order, from shortest to tallest. Post them on the wall to make a height museum.

Half as high (Hard). Fold your strip in half so it is half your height. Then, find things in the room that are half your height.

Levels: Easy, Medium (Hard)
Group size: 1 per project

Strawberries and Blueberries

Do you have more strawberries or more blueberries? Compare handfuls to find out.

Materials

Per project

construction paper
marker or crayon

To share

bowl of "strawberries" (red pompoms or plastic bottle caps), sized so that a child can hold 2-4 in one hand

bowl of "blueberries" (similar objects in blue)

glue stick
number stickers (optional)

1 Trace

Trace your hand on construction paper.

2 Reach for the red

Grab a handful of strawberries.

Put each strawberry on a finger you traced.

 Easy: How many fingers have a strawberry?
Medium: Do you have more fingers or more strawberries?

3 Add the blue

Take a blueberry for each empty finger. If you don't have any empty fingers, don't take any blueberries.

4 Glue them down

Glue your strawberries and blueberries.

(Optional) Count your strawberries and blueberries. Write the numbers on your paper or use number stickers.

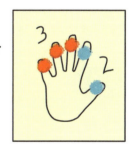

Variations

Shut your eyes and take five (Medium). Mix the strawberries and blueberries in one bowl. Shut your eyes, and count out five objects from the bowl. Then, arrange them on your traced hand.

Make it ten (Hard). Trace two hands. Take a handful of strawberries and match them with fingers. Then, take the rest in blue.

© 2014 TERC · Cambridge, MA

PROJECTS AND CRAFTS

Soaring Towers

Levels: Easy, Medium, Hard
Group size: 1–2 per tower

Stack blocks to build the highest tower you can. Then try to go even higher!

1 Build

Use some or all of your blocks to build the highest tower you can.

Your tower must stay up on its own.

How did you make your tower stable, so it doesn't fall over? What shapes are in your tower?

How many blue blocks in your tower?

What would happen if you put a sphere on top?

Materials

Per tower

varied blocks or small boxes

   Easy. 4-6

   Medium. 7-10

   Hard. 11-15

2 How high does it go?

Compare the tower to something nearby.

Does your tower go up to your knees? Does it go up to the top of the table?

3 Rebuild

Try again and see if you can build a higher tower. Use the same blocks or trade one of your blocks with someone else.

What if you turn the block long side up? Would that make your tower higher?

Your last tower was about as high as the chair. Is this one higher than the chair?

Variation

Towers over time (Hard). Cut a paper strip to the height of each tower you build. On each strip, write your name, the date, and the number of blocks you used in the tower. Did the highest tower use the most objects?

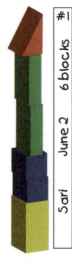

36   **PROJECTS AND CRAFTS**   © 2014 TERC • Cambridge, MA

Levels: Easy, Medium, Hard
Group size: 1 per book

Picture Book

Make and share a counting book.

1 Decide how high you'll count

Easy. Start at 1 and count to 4.
Medium. Start at 1 and count to 6.
Hard. Start at 1 and count to 12.

Materials

Per book
several sheets of construction paper, folded in half and stapled to make a book

To share
markers, stickers
number stickers (optional)

2 Choose a theme

 Should we use puppy stickers or kitten stickers?

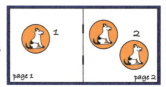

3 Put it on paper

Match a number of puppy stickers to the count: one on the first page, two on the second, and so on.

 What comes after two? Show me with your fingers. Let's count three puppy stickers.

4 Number it (optional)

Use a number sticker or write a number on each page to show how many.

 How many stickers on this page? Which sticker shows a '3'?

5 Read your book

Count the objects out loud as you show each page of your book.

Variations

Tell a story (Easy, Medium, Hard).
Make up a story to go along with your book.

Different ways to count (Hard). Count by 1¢ up to 10¢, start at 0 and count by 2s up to 10, or count back from 10 to 1.

Dimes and pennies (Hard). Start at 10¢ and count up by pennies.

I saw one puppy. Another one ran over and there were two!

PROJECTS AND CRAFTS

Fill It Up

Fill up a cup with pompoms. Challenge others to find how many are inside.

Levels: Easy, Medium, Hard
Group size: 2 or more (to make and trade cups)

1 Fill It Up

Count out objects to fill a cup. Keep the total a secret!

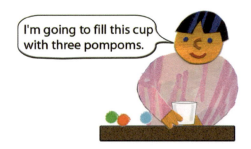

 You counted three pompoms. If you count them again will you get three?

2 Trade cups

Estimate how many are inside.

 Do you think there are any in the middle of the cup that you can't see?

3 Count

Take the objects out of the cup. Count to check your estimate.

Materials

Per project

clear plastic cup

To share

bowl of pompoms or other objects in different colors for filling cups

Easy. Objects sized so 3-4 fill the cup.

Medium. Objects sized so 5-6 fill the cup.

Hard. Objects sized so 10-12 fill the cup.

Variations

Change the shape (Hard). You'll need two clear plastic containers. Each container should be a different shape. Fill each with identical objects. Predict: Which has more inside? Why do you think so? Count to check.

Change the size (Hard). You'll need two clear plastic cups and objects that come in two sizes (e.g., large and small pompoms.) Fill each cup. Predict: Which cup has more inside? Why do you think so? Count to check.

38   PROJECTS AND CRAFTS   © 2014 TERC • Cambridge, MA

Levels: Easy, Medium, Hard
Group size: 1–2 per city

Cityscape

Building with boxes? Make a city and trace your cityscape.

Before beginning (Optional)

Clear off a section of a wall. Tape newsprint along the wall.

Materials

Per city

10–20 varied boxes or blocks

tape, markers, several sheets of newsprint (optional)

1 Build

Use your boxes and blocks to make a city.

Easy. Build in a pattern: tall building, short building, tall, short …

Medium. Build smaller to larger: each building is taller than the one before it.

Hard. Build by the numbers: 3 tall buildings, 3 short buildings, 3 tall, 3 short …

If you will be tracing your city, build against the newsprint on the wall.

> **Talk About** *If you turn the box sideways, is it taller or shorter? How many sides does the box have?*

2 Give a city tour

Tell about your cityscape.

> **Talk About** *Which parts are short? Which are tall? Which box is last?*

3 Trace and draw (optional)

Trace your city outline on newsprint. Remove the newsprint from the wall, and draw in the buildings.

Variations

Who lives there? (Easy, Medium). Find toy animals the right size to live in the buildings in your city.

Try it outside (Easy, Medium, Hard). Build your city along the sidewalk. Trace your city outline with chalk.

Toyscape (Easy, Medium). Line up a set of toy vehicles, people, or animals in order from shortest to tallest. Or, arrange them in a pattern based on height.

PROJECTS AND CRAFTS

Watch Me Grow

Levels: Easy, Medium, Hard
Group size: 1-2 per plant

How does your garden grow? Plant seeds and track how they grow over time.

Before beginning

Plant seeds or seedlings.

1 Draw (optional)

Look carefully at your plant each week. Draw what you see in your plant book.

2 Measure

Measure your plant with a straw each week. Line up a straw against the plant.

An adult helps cut the straw to the height of the plant. Tape the straw to graph paper, and mark the date.

Make sure to water your plant each week.

3 Observe

Easy. How many leaves on your plant? Did it have the same number last week?

Medium. How did your plant change this week? Is it taller?

Hard. Does your plant grow about the same amount each week? How can you tell?

Materials

Per plant
- seeds or seedling that grows quickly (e.g., grass, beans, avocado pit)
- potting soil and a pot
- a few straws
- piece of graph paper
- several sheets of blank paper, folded and stapled to make a book (optional)

To share
- tape and crayons or markers

For an adult
- scissors

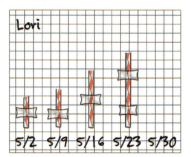

Variations

Pair up the plants (Hard). Grow two plants. Mark the pots so you can tell them apart. Record their growth on the same chart, using two different color straws. Are both plants the same height each week?

Change the conditions (Hard). Put one plant in the sun and one in the shade. Do they both grow the same amount?

Projects and Crafts

Level: Medium (Hard)
Group size: 1 per mask

Make a Mask

Fold, draw, and cut. Then, unfold to see your mask!

Materials

Per mask
piece of construction paper
marker
scissors
string

To share
crayons or markers

1 Fold
Fold a piece of paper in half.
Place it so that the fold is on the right.

2 Draw
Draw a shape that starts near the top of the fold and ends near the bottom.

3 Cut
Cut along the lines you drew.
Don't unfold the paper yet!

4 Predict
Talk About — *What will you get when you unfold the paper? How many points will be at the side?*

5 One, two three ... unfold!
Talk About — *Does it look like you predicted?*

Decorate your mask.

An adult helps cut out eye holes, punch holes, and put in string so you can wear your mask.

Variations

Start from the whole (Hard). Use folded paper to make a heart, tree, or other shape that is the same on both sides. Plan in advance so that when you cut and unfold, you get the shape you want.

Quarter it (Hard). Fold the paper in half and in half again. Draw an outline of one quarter of a flower, snowflake, or design. Predict what you'll see when you cut along the line you drew. Then, cut and unfold.

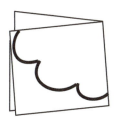

Projects and Crafts

Chain Challenges

Making paper chain decorations? Use them to measure.

Levels: Easy, Medium, Hard
Group size: 1 per chain

1 Link up!

Make a chain with your links.

2 How long is your chain?

Find something the same size as your chain.

 Does your chain fit across the chair? Does it go up to your knees?

3 Predict

 Easy. *If we tape two chains together, how long will the big chain be? Show me with your arms.*

Medium. *If we tape four chains together, will they go up to your waist?*

Hard. *If we tape ten chains together, will they be as tall as you?*

4 Tape your chains together and test your predictions

Materials

Per chain

strips of colored paper about 1" wide and 8" long:

Easy. 5
Medium. 10
Hard. 15

glue stick

To share

one or two rolls of tape

Variations

Pattern party (Easy, Medium, Hard). Alternate two link colors or use your links to make a different pattern.

Made to measure (Medium, Hard). Make a chain of a certain size: as long as your foot, as tall as you are, or the length of the room.

Double trouble (Hard). If you use links twice as long, how long will your chain be? Try it and see!

Trail of tens (Hard). Everyone makes a chain of 10 links. Each person uses a different color. Tape them together and count by 10s to find out how many chain links in all.

Levels: Easy, Medium, Hard
Group size: 1–2 per tower

Towering Toothpicks

Select, connect, and build a tower that is tall and strong!

Materials

Per tower

about 100 grapes, gumdrops, or stale mini-marshmallows (more if you're eating while building)

about 200 toothpicks

Medium. Small potato

Hard. Large potato

1 Explore

Use your materials to experiment.

 Do you have any triangles in your tower?

How many squares can you find in you tower?

How could you make your tower taller?

2 Build, test, and revise

Use the materials to build …

Easy. … the tallest tower you can.

Medium. … a tower that holds up a small potato.

Hard. … a tower that holds up a large potato.

 How can you make your tower stronger? What shapes hold together well?

3 Show and tell (optional)

Demonstrate your project.

 What shapes did you use? How did you make your tower strong?

Variations

Build big (Easy). Use craft sticks and large marshmallows. Try to build the tallest tower you can.

Try it with 20 (Hard). Make the tallest tower you can with 20 toothpicks and 10 gumdrops. If several people are building towers, compare sizes and shapes.

Build a bridge (Hard). Use toothpicks to build a bridge that a toy car can ride under or over.

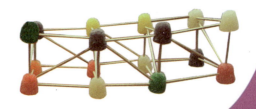

PROJECTS AND CRAFTS

Levels: Medium (Hard)

Make It Morse

String up your first initial in a necklace using Morse code.

1 Make it in Morse code

Make your first initial in Morse code. Use round beads for the dots and long, thin beads for dashes. String up the beads.

Materials

Per necklace

several beads, long and thin in one color

several beads, small and round in a different color

string or lace for beading

Morse Code Chart (p. 45)

2 Morse code show and tell

 How many round beads did you use? How many long ones? How many in all?

Which of us used the most beads?

3 Make a necklace

Knot your string and wear your necklace!

Variations

Make it for me (Hard). Work with a partner. Give your partner step-by-step directions for making a Morse code chain with your first initial. Then, switch roles so your partner gives you directions.

Morse code mix-up (Hard). Three or four people with different first initials put their chains out, mix them up, and take one. Don't take your own. If you can figure out the letter on the chain, return it to the owner. Otherwise, ask the owner to claim it.

Name it (Hard). String up your first name or your initials in Morse code.

44 PROJECTS AND CRAFTS © 2014 TERC • Cambridge, MA

Group size: 1 per necklace

Morse Code Chart

A	→	•−	N	→	−•
B	→	−•••	O	→	−−−
C	→	−•−•	P	→	•−−•
D	→	−••	Q	→	−−•−
E	→	•	R	→	•−•
F	→	••−•	S	→	•••
G	→	−−•	T	→	−
H	→	••••	U	→	••−
I	→	••	V	→	•••−
J	→	•−−−	W	→	•−−
K	→	−•−	X	→	−••−
L	→	•−••	Y	→	−•−−
M	→	−−	Z	→	−−••

Projects and Crafts 45

Hands on the Wall

Make a quilt of handprints to hang on the wall.

Levels: Easy, Medium, Hard

Group size:
Easy. 2–4
Medium. 2–6
Hard. 2–9

1 Share out the quilt pieces

 We have six quilt pieces and three people. If we share out the quilt pieces, how many does each person get? Are there any extras?

Divide up the quilt pieces, and put any extras aside for now.

2 Dip in

Dip your hand in a paint tray, and make a handprint on each of your quilt pieces.

3 Decorate any extra quilt pieces (optional)

If you have any left over, decorate them together. You will use them in the quilt, too.

4 Lay out the quilt pieces

Lay the cardboard on the floor or a table. Work together to arrange the quilt pieces to cover the cardboard.

 How many quilt pieces fit across?
How many fit down?
Are the quilt pieces the same size all around?

5 Glue

Once your quilt pieces are in place on the cardboard, glue them down.

Hang up your quilt when the glue is dry.

Materials

Per quilt

piece of 18" × 24" cardboard or newsprint

tempera paint in a few colors, each in a tray for dipping hands

glue stick

"quilt pieces" (rectangles made from light-colored construction paper):

  Easy. 4 rectangles (9" × 12")

  Medium. 6 rectangles (6" × 12")

  Hard. 9 rectangles (6" × 8")

Variations

Trace and decorate (Easy, Medium, Hard). Instead of dipping your hand in paint, trace it, and then decorate the outline of your hand.

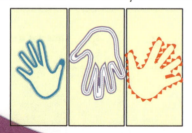

Giant size (Hard). Use a 24" × 36" piece of cardboard or newsprint and eighteen 6" × 9" rectangles.

PROJECTS AND CRAFTS

Levels: Easy (Medium, Hard)
Group size: 2 per pattern

Copy Cat

I'll make a pattern. Copy it if you can!

Materials

Per 4 patterns

blank shapes (pp. 61-67)
glue stick
plain white paper

1 Start the pattern

One person chooses three shapes to start.

 What shapes did you pick? What colors are they?

Glue them down in a row.

2 Copy

The other person repeats the pattern.

 What shapes will you use? Which one goes first? Which goes second?

3 Switch roles

Repeat steps 1-2. This time, the person who copied starts the pattern, and the other person copies.

Variations

Copy two (Easy). Repeat two shapes instead of three.

Include orientation (Hard). Use different orientations for the shapes in your pattern.

Pattern path (Easy, Medium, Hard). Instead of gluing your pattern on paper, tack it to the wall or along the floor to form a path from one part of the room to another. If you have wooden or plastic shapes, you can use those in place of paper shapes.

Copy color and shape (Hard). An adult photocopies blank shapes (pp. 69-70) three times: each on a different color paper. Use color *and* shape to make your patterns.

Outdoor trail (Easy, Medium, Hard). Make a pattern path on the sidewalk by tracing shapes with chalk.

PROJECTS AND CRAFTS

Levels: Hard (Easy, Medium)

Say It with Shapes

Make pattern poems with shapes and words.

1 Create the first line

Choose 3-4 words to use for your first line. Place them on plain white paper.

 What is your poem about? What shapes are you using?

2 Continue the pattern: same shapes, different words

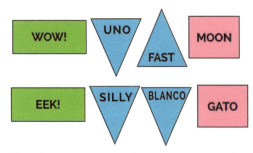

Repeat the shape pattern at least twice to make a poem.

3 Edit and glue

When you are satisfied with your poem, glue it down on paper.

4 Share your poem

 What shape comes after the rectangle in your poem? What is the color pattern in your poem? What is the shape pattern?

Materials

Per 4 pattern poems
- one word set (pp. 51-55)
- glue stick
- plain white paper

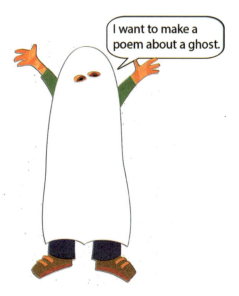

48   PROJECTS AND CRAFTS   © 2014 TERC • Cambridge, MA

Group size: 1 per pattern poem

Say It with Shapes (cont'd)

Variations

Grow a poem (Hard). Make a pattern that "grows" each line. The poem could grow by adding a shape each line.

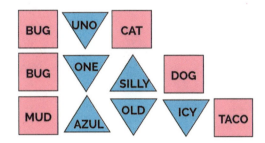

Or, the poem could grow symmetrically:

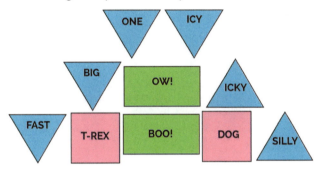

Letter it (Medium). Combine letters (pp. 57-59) and words (pp. 51-55) to make a visual poem.

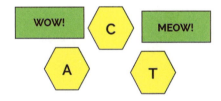

Shape and color patterns (Easy, Medium, Hard). Use blank shapes (pp. 61-70) to make a pattern.

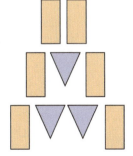

© 2014 TERC • Cambridge, MA

PROJECTS AND CRAFTS   49

Levels: Hard (Easy, Medium)
Group size: 1 per picture poem

Picture Poems

Make a picture with shapes and words.

1 Decide on a theme

 What will your poem be about?

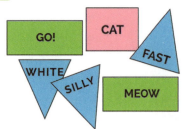

Materials

Per 4 picture poems
one word set (p. 51-55)
glue sticks
plain white paper

2 Make a picture

Choose words that fit your theme and arrange them in a picture or a design.

3 Share your poem

 How many squares did you use in your picture? What shape is above the rectangle?

Variations

Holiday cards (Hard). Use words and an image to fit the season.

Shape pictures (Easy, Medium, Hard). Combine blank shapes (pp. 61-69) to make a picture.

Letter it (Medium, Hard). Make a pattern or picture with letters (p. 57-59) in your name.

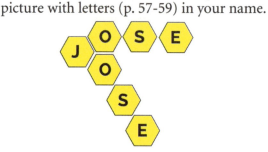

Or, combine shapes, letters, stickers, and drawing.

50   **PROJECTS AND CRAFTS**   © 2014 TERC • Cambridge, MA

Nouns

CAT	T-REX	TACO
DOG	MUD	TRUCK
BUG	SUN	NIGHT
GATO	MOON	DAY

✂ Cut out along lines. Find more shapes on pp. 111-133.

© 2014 TERC • Cambridge, MA

PROJECTS AND CRAFTS   51

Exclamations

EEK!	GO!
OOPS!	OH!
MEOW!	WOOF!
YIKES!	WHEE!
BOO!	YUM!
¡SÍ!	¡BRAVO!

Cut out along lines. Find more shapes on pp. 111-133.

© 2014 TERC • Cambridge, MA

PROJECTS AND CRAFTS

Adjectives

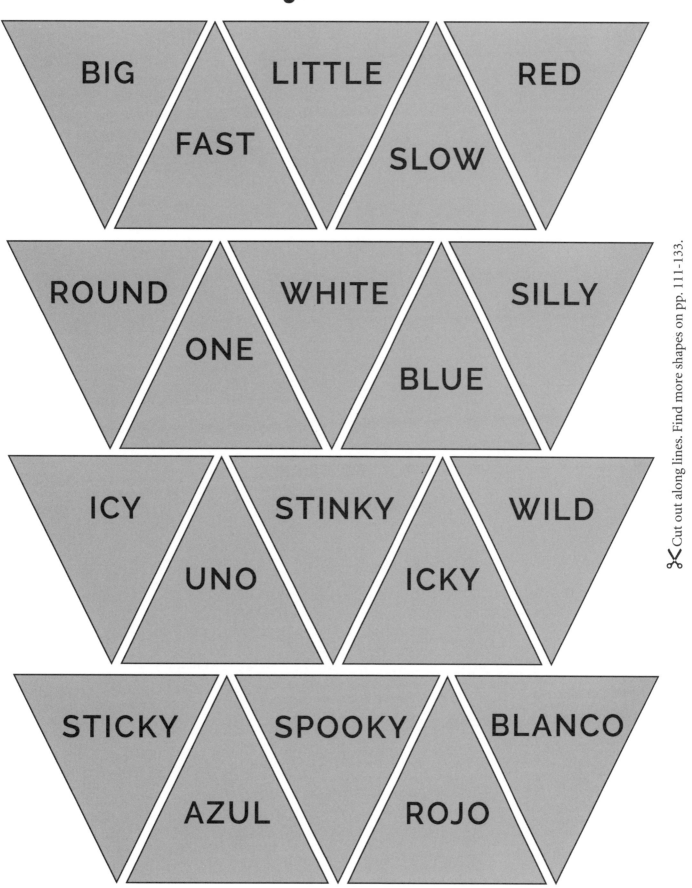

✂ Cut out along lines. Find more shapes on pp. 111-133.

© 2014 TERC • Cambridge, MA   Projects and Crafts   55

Letters

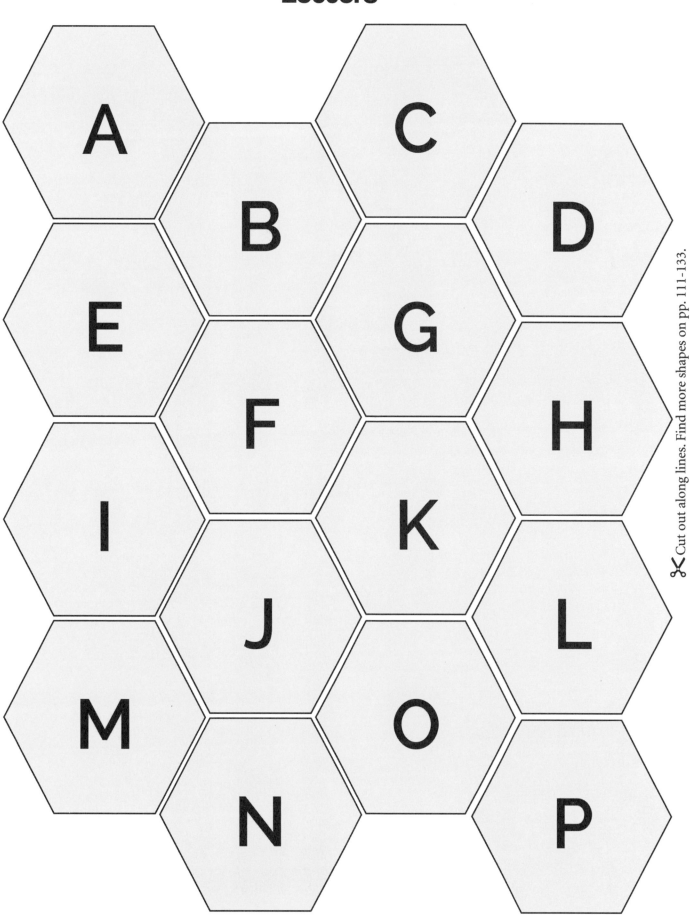

Cut out along lines. Find more shapes on pp. 111-133.

© 2014 TERC • Cambridge, MA

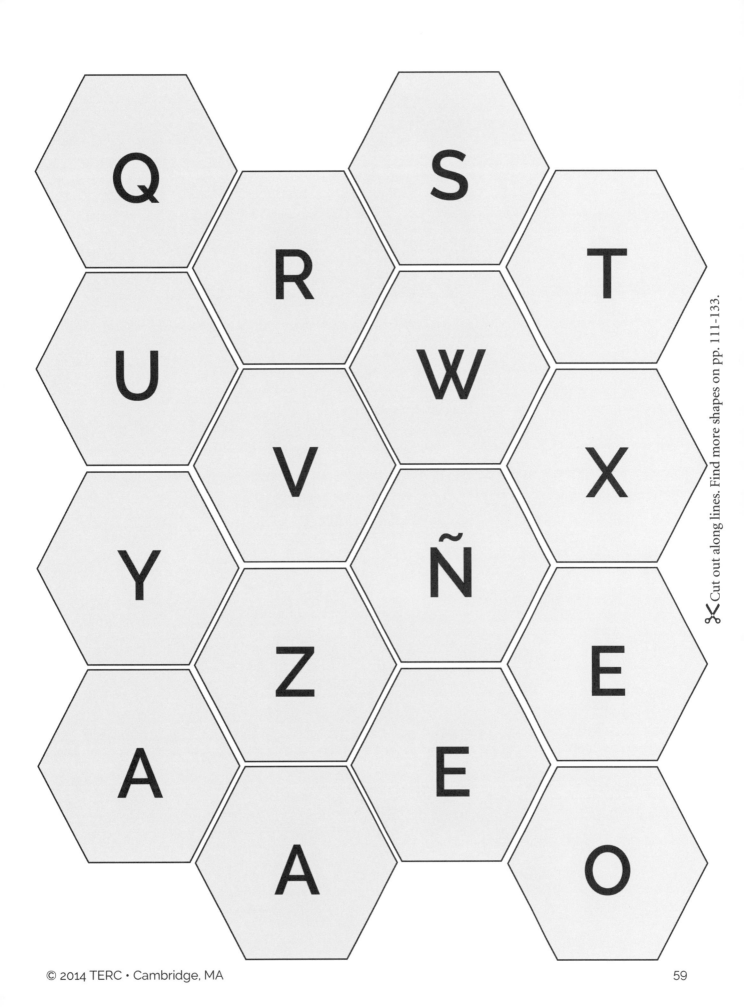

Cut out along lines. Find more shapes on pp. 111-133.

© 2014 TERC • Cambridge, MA

59

Squares—blank or fill in with nouns

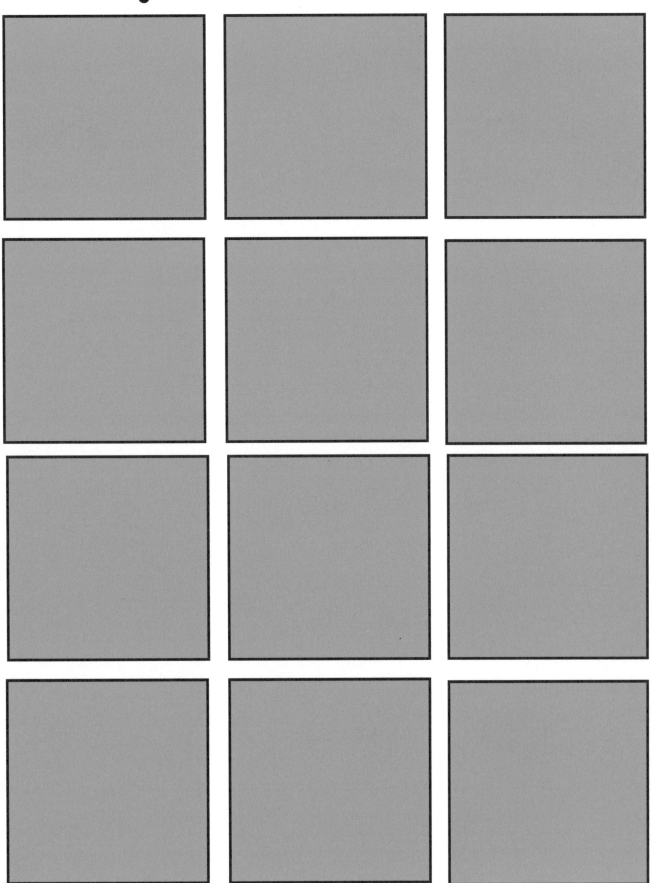

✂ Cut out along lines. Find more shapes on pp. 111-133.

Rectangles—blank or fill in with exclamations

Cut out along lines. Find more shapes on pp. 111-133.

© 2014 TERC • Cambridge, MA

63

Triangles—blank or fill in with adjectives

✂ Cut out along lines. Find more shapes on pp. 111-133.

© 2014 TERC • Cambridge, MA

Hexagons—blank or fill in with letters

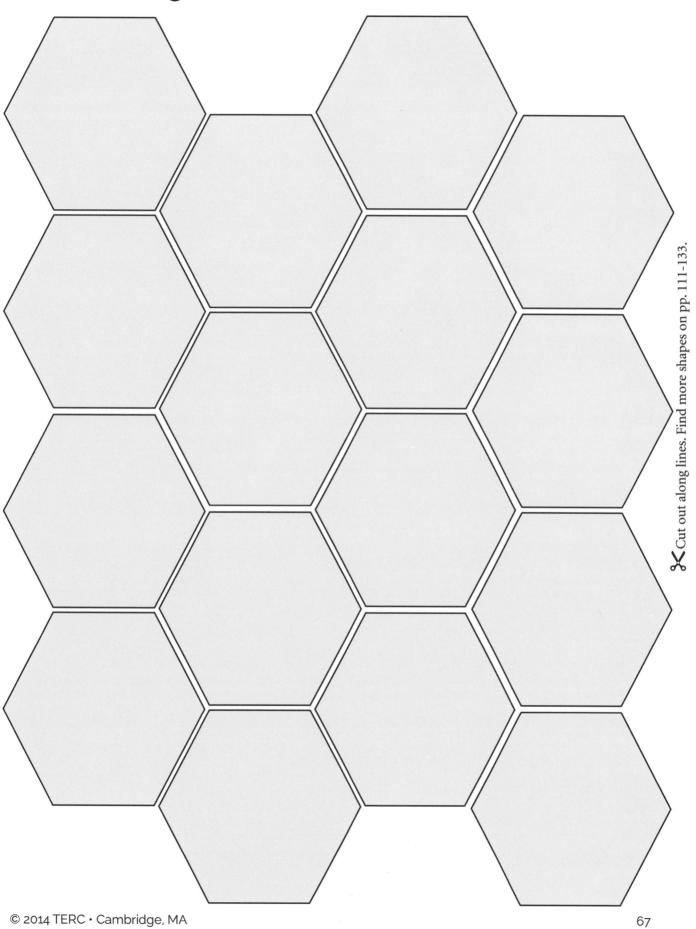

✂ Cut out along lines. Find more shapes on pp. 111-133.

Blank Shapes

Photocoy on colored papaer and cut out. ✄

© 2014 TERC • Cambridge, MA

PROJECTS AND CRAFTS 69

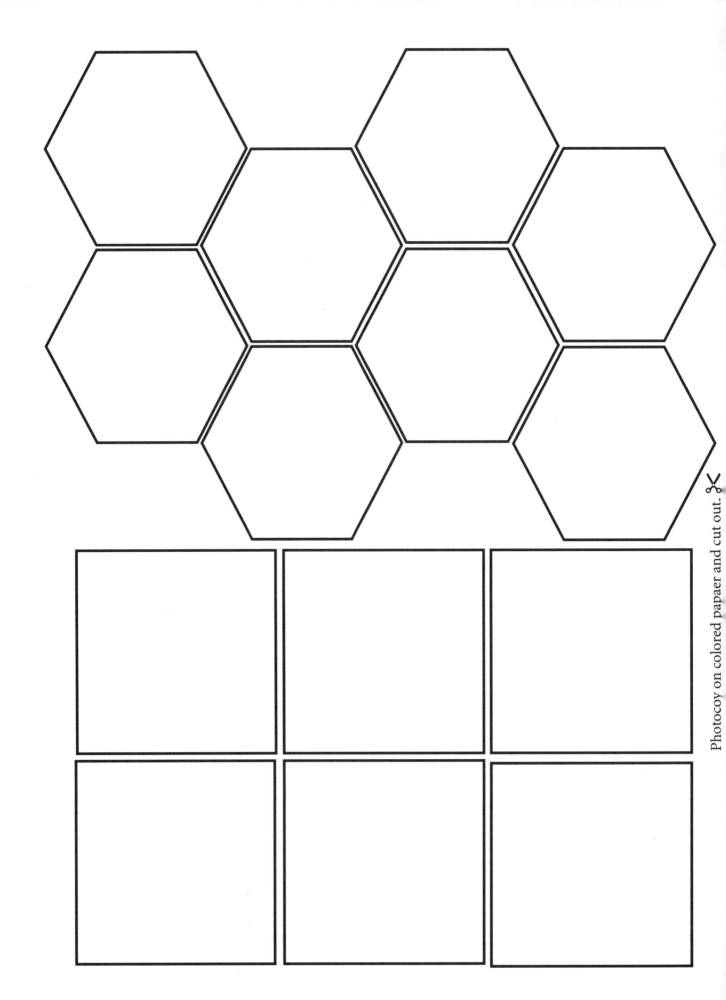

Food and Water

Ideas to investigate and games to play in the kitchen, at snack time, or around water.

Contents

Fill It or Spill It 72
Snack Station 73
Water Wonders 74
One for You, One for Me 75
Pretend Picnic 76
What's Inside? 77
In the Bag 78
My Height in Boxes 79
The Counting Chef 80

Food and Water activities in other sections

Projects and Crafts

Strawberries and Blueberries 35
Cityscape 39
Watch me Grow 40
Towering Toothpicks 43

Good for Groups

Quick Questions 85

© 2014 TERC • Cambridge, MA

Fill It or Spill It

Fill up the bowl with water, one cup at a time.

Levels: Easy, Medium, Hard

Group size: small enough so everyone gets a turn to pour

1 Compare sizes

Pass around the cup and the bowl.

Easy. Which is wider, the cup or the bowl? Which is taller? Which could hold more water?

Medium. Which do you think can hold more water: the cup or the bowl? Why do you think so?

Hard. How many cups of water do you predict will fill the bowl?

Materials

Per group

large bowl

cup sized so that:

    Easy. Up to 4 cups of liquid fill the bowl.

    Medium. Up to 6 cups of liquid fill the bowl.

    Hard. Up to 12 cups of liquid fill the bowl.

water source (e.g., faucet or bucket of water)

2 Fill and count

Take turns adding a cup of water to the bowl. Keep going until it's full.

Easy. One cup ... two cups ... what number comes next? Is the bowl full yet?

Medium. Do you think we have room for another cupful? ... two more cupsful?

Hard. We filled the bowl with nine cups. Is is that more or less than you predicted?

Variations

Mix it up (Easy, Medium, Hard). For experience with different sizes and shapes, use varied containers (e.g., tall and thin, short and wide) instead of a cup and bowl.

Fill it a different way (Easy, Medium, Hard). Fill with sand, rice, or beans instead of water.

Count around (Hard). Try this with at least two people. Gather in a circle. On your turn, pour a cupful into the bowl and say the next number in the sequence. For extra challenge, pour two cups and count around by 2s.

72   FOOD AND WATER

Levels: Easy, Medium, Hard

Group size: 2 per order (1 server and 1 to be served)

Snack Station

Let me take your order. How many do you want?

Materials

Per pair

food to count out, such as baby carrots or crackers

1 Set the limit

Decide on the maximum amount per order.

Easy. Up to 4 items.

Medium. Up to 6 items.

Hard. Up to 12 items.

I can eat three crackers.

2 Pair up

Decide who will serve first and who will order first.

3 Order your snack

Say how many items you want.

Talk About: *How hungry are you? Do you think you could eat just one cracker? Could you eat five?*

The server counts out your order.

1, 2, 3 crackers.

4 Switch roles.

If you ordered last time, this time you serve.

Variations

Small, medium, large (Hard). Decide together how many items in a "small" serving, a "medium" serving, and a "large" serving. Then, one person orders a small, medium, or large, and the other counts it out.

One cracker is a small order, three is medium, and six is large.

Fantasy feast (Easy, Medium, Hard). "Feed" toy animals, dolls, or vehicles as many items as they request.

Pour it (Hard). You'll need food that can be poured (e.g., cereal, yoghurt), a measuring cup, bowls, and spoons. Decide on measurements for each serving size. For instance, small is 1/4 cup, medium is 1/2 cup, and large is 3/4 cup. Order and serve.

Food and Water

Water Wonders

Levels: Easy, Medium, Hard
Group size: 1–3 per set of containers

Which holds more water: the tall, thin container or the short, wide one? Compare to find out.

1 Predict which holds the most

Put the containers in order from the one you think holds the least to the one you think holds the most.

 Why do you think this short, thin container holds the least?

2 Fill and pour

Fill the container you think is the smallest. Then pour the water into another container. If it's really the smallest, the water won't completely fill up any other container.

 Does the water fill up the container? Is there a lot of room left?
Does the tallest container hold the most water?

Materials

Per group

plastic containers, cups, or bowls of different shapes and sizes (e.g., short and round, cone-shaped, tall and thin)

Easy. 2 containers.
Medium. 3 containers
Hard. 4-5 containers.

water source (e.g., faucet or bucket of water)

Variations

Fill it a different way (Easy, Medium, Hard).
Fill with sand, rice, or beans instead of water.

One dimension at a time (Easy, Medium).
Line up the containers from shortest to tallest. Then line them up from narrowest to widest. What other ways can you find to line them up?

Cereal box lineup (Easy, Medium, Hard).
Instead of containers, use cracker, tea, or cereal boxes. Line them up by height, width, or another characteristic (e.g., number of "A"s on the front of the box).

Levels: Easy, Medium, Hard

Group size: 2
(see Variations for more)

One for You, One for Me

Snack time? Deal out the food to make sure everyone gets an equal share.

Materials

Per pair

food that comes in pieces, such as baby carrots or apple slices, arranged on a plate so everyone can see how many

Easy. 3-4 pieces.
Medium. 5-6 pieces.
Hard. 7-12 pieces.

1 Predict

 Are there more people or more carrots? Could each person have one? How do you know? Could each person have two?

2 Each one takes one

Each person takes one item from the plate.

 Are there any left over? Can everyone have another?

3 Keep taking one

Repeat step 2 until the plate is empty or only one item remains.

 Do you both have the same amount? How do you know?

4 Eat!

If there is an extra left on the plate, divide it in half or save it for another time.

Variations

Feed a group (Medium, Hard). Try this with three, four, or five people sharing food.

Place settings (Easy, Medium). Set out a pile of plates, forks, and napkins. Predict if you have enough for everyone and then lay the table to check.

Feed your toys (Easy, Medium, Hard). "Help" two or more toy animals, dolls, or vehicles share out food so each one gets the same amount.

© 2014 TERC • Cambridge, MA

FOOD AND WATER

75

Pretend Picnic

Levels: Easy, Medium, Hard
Group size: small enough so everyone has a turn to remember everything at the picnic

What are you taking to the picnic? Remember what everyone else is bringing too.

1 Start with a 'one'

The first person announces one thing to bring on a pretend picnic.

2 Contribute a 'two'

The next person repeats what the first is bringing and contributes two more.

3 Keep going as long as you can!

Easy. Up to 3.

Medium. Up to 5.

Hard. Up to about 10.

 What are we taking two of? What number comes after four?

Variations

Picture picnic (Easy, Medium, Hard). Use pictures, stickers, or real or toy food to help you remember what you're bringing on the picnic. You might reach a higher count that way!

Alphabet picnic (Hard). Follow the alphabet and the counting sequence as you decide what to bring on your picnic.

Backwards picnic (Hard). Start with six items and count back each turn. For extra challenge, start at ten.

Picnic in time (Hard). Follow the clock as you tell what you did on the picnic. For instance, "At 1:00 in the afternoon, I saw 1 butterfly. At 2:00, I saw 2 frogs. ..."

Levels: Medium, Hard (Easy)

Group size: small enough so everyone has a chance to predict and count

What's Inside?

How many seeds in this apple? Predict, count, and eat!

Before beginning
Cut open the fruit so that the seeds are visible.

Materials

Per group

Medium. Fruit with up to 10 seeds (e.g., apple, orange).

Hard. Fruit with up to 15 seeds (e.g., pear, melon slice).

For an adult knife

1 Predict the number of seeds

 Are there more than two seeds? Are there more seeds than you have fingers on one hand?

2 Count

Remove, arrange, and count the seeds.

 Let's count together to find how many seeds.

3 Compare estimate and count

 "I thought apples had five seeds."

Were there fewer or more than you predicted?

4 Eat!

Variations

Count together (Easy). Divide up the seeds so everyone counts 2-4. An adult helps find the total.

Section selection (Medium, Hard). Predict how many sections in an orange. Then peel, count, and eat.

Peas in a pod (Medium, Hard). Try this with different kinds of peas or green beans. Make a prediction, then open the pod or hold it up to the light to count.

Edible explorations (Medium, Hard). Gather several types of fruit. Predict how many seeds in each, then an adult cuts. Do the larger fruits always have more seeds?

Apple graph (Hard). Track and chart how many seeds in each apple you eat for a month.

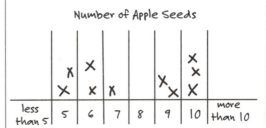

© 2014 TERC • Cambridge, MA

FOOD AND WATER

In the Bag

What's in the bag? Try to identify objects by shape.

Levels: Easy, Medium, Hard

Group size: small enough so everyone has a turn to hold the bag

Before beginning

Secretly hide the fruit or vegetable in the bag.

1 Take turns feeling the bag

Describe what you can feel through the bag. Don't put your hand inside the bag.

Easy. *I feel something big and round. Let's feel it together.*
Medium. *Is it round? long? pointy?*
Hard. *How can you describe the shape?*

Materials

Per group

familiar fruit

opaque, cloth bag large enough to hold the fruit or vegetable

For an adult

knife

2 What could it be?

Everyone makes a prediction.

3 Remove the food

Take out the fruit or vegetable and pass it around.

When this mango was in the bag, we noticed that it had a wide part and a narrow part. Show us the wide part.

4 Cut and eat

We noticed a prickly, pointy part of the pineapple. Do we eat that part? Why or why not?

Variations

Numbers in the bag (Easy, Medium, Hard). One person puts several objects in the bag. Others say how many they think are inside.

Shapes in the bag (Easy, Medium, Hard). One person puts an object with a geometric shape (e.g., ball, small box) in the bag. Others describe what they feel and predict the shape.

Ask about it (Medium, Hard). One person puts an object in the bag. Others pass around the bag, and take turns asking yes-or-no questions to figure out what it is.

Levels: Easy, Medium, Hard
Group size: any

My Height in Boxes

How tall are you in cereal boxes? in juice boxes?
Save them, stack them, and measure yourself!

1 Predict

Take a look at the boxes and compare them to your size:

Easy. *Are you taller than this cereal box?*

Medium. *If we put two cereal boxes on top of each other, would they be as tall as you are?*

Hard. *If you stack up cereal boxes, how many would you need to make a tower as high as you are?*

Materials

Per child

enough identical boxes to make a stack to measure height

2 Stack

Stack up the boxes to about your height. They might end up a little shorter or taller than you are.

Is the stack taller than you are? How do you know?
Are all the boxes turned in the same way, so they're as tall as they can be?

3 Count

About how many boxes tall are you?
How does that compare with your prediction?

Variations

Measure around the room (Easy, Medium, Hard). Measure in cereal boxes: the height of a chair or small table (Easy), the width of the room (Medium for a small room), or the length of a hallway or large room (Hard).

Find something (Hard). Everyone gets a cereal box, a juice box, and a cracker box. Find something in the room that is as wide as each box and something in the room that is as tall as each box.

Compare measurements (Hard). Predict: if you measure yourself using smaller boxes (e.g., juice or cracker boxes) will you need more or fewer than the number of cereal boxes? Try it and see.

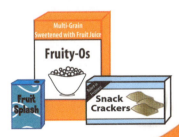

© 2014 TERC • Cambridge, MA

FOOD AND WATER

Levels: Easy, Medium, Hard

Group size: small enough so everyone has a turn to measure

The Counting Chef

Get everyone measuring and mixing to make trail mix, bubble soap, playdough, or lemonade.

1 Talk through the measures and ingredients

Which one is the cup measure? Show me the tablespoon.

Playdough
2 cups flour
2 cups warm water
1 cup salt
2 T vegetable oil
2 T cream of tartar
food coloring, optional

Materials

a recipe calling for whole number amounts

Easy. Up to 2 "scoops" per ingredient.

Medium. Up to 4 "scoops" per ingredient.

Hard. Choose an "Easy" or "Medium" recipe to double.

individual measures (e.g., cup scoops or dry measuring cups)

related ingredients and supplies

2 Measure it out

Easy. *We need two cups of flour. Let's count out one cup at a time.*

Medium. *Can you measure out two cups of water?*

Hard. *The recipe calls for two tablespoons of vegetable oil. We're making twice the recipe. How many tablespoons do we need?*

3 Make the recipe

Enjoy the results!

Variations

Measure up (Hard). Use a measuring cup instead of individual cup measures.

Include halves (Hard). Try a recipe that involves 1/2 cup or 1 1/2 cups.

Smoothie cookbook (Medium, Hard). Each day, mix up a different combination of juice, fruit, and yogurt. Write down each recipe with stickers, pictures, or words. Put the best results in a cookbook.

Good for Groups

Group games, party favorites, icebreakers, and circle time activities for indoors and out.

Contents

Catch the Beat	82
Group Up	83
Stand and Vote	84
Quick Questions	85
Toy Store	86
Who's Here?	88
Line Up	89
Treasure Hunt	90

Good for Groups in other sections

Games

Secret Card (Variation, Secret person)	6

Projects and Crafts

Chain Challenges	42

Food and Water

Fill It or Spill It (Variation, Count around)	72
One for You, One for Me (Variation, Feed a group)	75
Pretend Picnic	76
In the Bag	78

Anytime, Anywhere

Words on the Wall (Variation, Play with a group)	92
Seeing Shapes (Variation, Play with a group)	96
Take Two (Variation, Take five)	98

© 2014 TERC • Cambridge, MA

Catch the Beat

Follow a rhythm pattern together and follow it.

Levels: Easy, Medium, Hard
Group size: 2 or more

1 Start a rhythm pattern with two beats

For instance, clap your hands, jump, clap, jump …

Materials

none

2 Keep repeating until everyone is following along

 What comes after "jump?"

3 Mix it up

Easy. Take turns doing a different part of the pattern. One person claps, the next jumps, the next claps, etc.

Medium. Double the beats: two claps, two jumps, two claps, two jumps …

Hard. Half the group keeps the beat; the other half doubles it.

 Do we jump as many times as we clap? How do you know?

Variations

More beats (Easy, Medium, Hard). Use a rhythm pattern with three or four beats.

Draw the beat (Hard). Half the group keeps the beat. The other half draws a visual pattern to show the beats they hear. Then, switch roles.

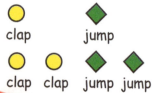

Play the beat (Easy, Medium, Hard). Play drums, shakers, or other rhythm instruments to mark the beats.

Bark to the beat (Easy, Medium, Hard). Mark the beat with animal sounds and actions. For instance, bark like a dog, roar like a lion, and stretch like a cat.

Levels: Easy, Medium, Hard

Group size:
Easy. 2-4
Medium. 5-6
Hard. 7-12

Group Up

Find everyone with the same number of shoelaces as you, and form a group together.

Materials

none

1 Announce something to match

Choose a number characteristic that everyone can see and compare, such as number of shoelaces, ponytails, or shirt buttons.

How many shoelaces are you wearing? Do you think anyone is wearing the same number as you?

"I have two shoelaces."

2 Group up

Stand in a group with everyone who has the same number as you.

How many shoelaces do Zara and Marcus have? Do you have the same number?

3 Compare

Are more of us in the two-shoelace group or the no-shoelace group? Why do you think we don't have a one-shoelace group?

Variations

Measure it (Hard). Group up with everyone about as tall as you. Or, group up with everyone whose hair is about as long as yours.

"We're the same height."

Scatter (Easy, Medium, Hard). Everyone with the same number goes into a different part of the room. For instance, if you have 0 ponytails, stand by the windows. If you have 2, stand by the door.

Group up by color (Easy). Group up with everyone wearing the same color shoes as you.

Match your name (Hard). Group up with everyone who has the same number of first name letters as you.

GOOD FOR GROUPS

Stand and Vote

Vote with your feet! Get in line to show your vote.

Levels: Easy, Medium, Hard

Group size:
Easy. 3-4
Medium. 5-6
Hard. 7-12

1 Pose a question with two clear choices

Materials

none

2 Vote with your feet

Form two lines, one for each choice.

Get in the line that matches your vote.

3 Compare votes

 How can we tell which choice got more votes? If you match up people in each line are there any extras?

4 Decide what's next

 Should we go with the majority? What should we do if there is a tie?

Go outside   Stay inside

Variations

More choices (Medium, Hard). Try this with 7-12 people. Pose a question with three answers and get into three lines.

Complex questions (Hard). Line up by categories that involve "and," "not," and "or." For instance, stand in the first line if you are wearing red or blue; in the second if you are not wearing red or blue.

84   GOOD FOR GROUPS   © 2014 TERC • Cambridge, MA

Levels: Easy, Medium, Hard

Group size:
Easy. 3-4
Medium. 5-6
Hard. 7-12

Quick Questions

Answer a quick question to break the ice.

Materials

Per group

large piece of paper, marker, and stickers for making the chart

dot stickers

Before beginning

Come up with a multiple choice question everyone will enjoy answering. The choices should be easy to show with stickers or pictures.

Write the question at the top of a large sheet of paper. Put the choices at the bottom.

1 Predict

Which of these fruits do you like the best? Do you think your favorite will be the most popular?

Would it be fair if some people voted more than once? Why or why not?

2 Everyone answers

Use dot stickers to show your answer.

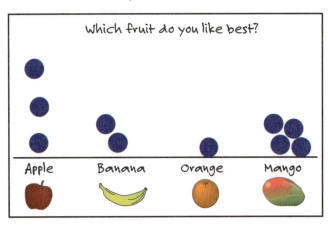

3 Look over the results

How many people chose "mango"?
How many more people chose "apple" than "banana?"
Which answer got the most dots?

Variation

Yes or no (Easy). Use a question with just two answers. For instance, "Do you like dinosaurs?"

© 2014 TERC · Cambridge. MA

GOOD FOR GROUPS

Toy Store

Buy and sell with play money and toys.

Before beginning

Sellers arrange their wares and attach price tags.

 What price tag will you put on that stuffed animal? Will you sell everything for the same amount?

Shoppers check how much money they have.

 How many bills do you have? How much play money do you have in all?

Materials

Per Shopper

play money

Easy. 5 $1 bills.

Medium. 10 $1 bills.

Hard. 5 $1 bills, 3 $2 bills, and 2 $3 bills.

shopping bag or basket for purchases

Per Seller

price tags (make from sticky notes)

Easy. Several $1 price tags.

Medium. Several $1 and $2 price tags.

Hard. Several $1, $2, and $3 price tags.

objects to "sell" (toy food, cars, animals)

Group size: at least 2 Shoppers and 2 Sellers

Toy Store, cont'd

"I started out with $10 and I bought a truck for $3."

1 Shop and sell

Shoppers circulate among the Sellers and make their purchases.

 How much have you spent so far? How much do you have left?
How much money will you have if you sell everything?
Do you have enough to buy two horses for $2 each? How do you know?

2 Switch roles

Today or another day, all the Sellers take a turn as Shoppers, and Shoppers have a chance to buy.

3 Return purchases and unsold goods

When you're done, put everything back for another day.

Variations

Write it yourself (Medium, Hard). Write the numbers on price tags and bills yourself.

Buy a set (Easy, Medium). Sell items that make up sets (e.g., fork, spoon, knife; individual puzzle pieces). If items cost $1 each, can you buy the whole set?

Go to the bank (Hard). One player is the Banker. The Banker gets a set of $1 bills and makes change when asked.

End of season sale (Hard). Decide on a discount, such as buy one get one free, half off, or $1 off for items that cost $2 or more. You might need a 50¢ coin or bill.

© 2014 TERC • Cambridge, MA

GOOD FOR GROUPS 87

Who's Here?

Is anyone missing today? Keep track of your group.

Levels: Easy, Medium, Hard

Group size:
Easy. 3-4
Medium. 5-6
Hard. 7-12

Before beginning
Make sure everyone knows the total number in the group.

Materials

none

1 How many are here today?

Get in a circle and count off to find out.

 We counted starting with Kiara and got to five. If we start with Enzo this time, do you think we'll also count five?.

Here ...

2 How many are missing?

Name and count those who aren't here.

... missing

3 Does it add up?

 Five of us are here. If we count the two missing, does that come out to seven? How can we check?

Variation

Five in a row (Hard). Keep an attendance record for five sessions. Explore the data. On which day were the most people absent? On how many days was everyone here?

Three kids were out Friday because it was a holiday.

Date	Present	Absent	Total
Monday, October 1	5	2	7
Thursday, October 2	6	1	7
Wednesday, October 3	6	1	7
Thursday, October 4	7	0	7
Friday, October 5	4	3	7

Levels: Easy, Medium, Hard

Group size:
Easy. 3-4
Medium. 5-6
Hard. 7-12

Line Up

Get moving when you're waiting in line.

Before beginning
Choose a size or number characteristic everyone can see and compare, such as height or number of pockets.

Materials

none

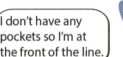
I don't have any pockets so I'm at the front of the line.

1 Predict

If we line up by number of shirt pockets you are wearing, do you think you'll be at the start, middle, or end of the line?

Take a look around you and make a prediction.

2 Line up

Make comparisons and stand in order. If two people have the same measurement or number, they stand one behind the other.

Who is taller? Let's stand back to back to check. Let's match up our pockets and see who has more.

3 Are we in order?

Check and change places if needed. The last person in line chooses how to line up next time.

Variations

Me first (Hard). Decide how to line up so that you're at the head of the line. For instance, if you're the tallest, you could ask everyone to line up from tallest to shortest.

Ask and line up (Hard). Line up by something you can ask about, for instance, by birth month, by birth year, or by number of pets.

I was born in May and Sam was born in June. Who goes first?

© 2014 TERC • Cambridge, MA

GOOD FOR GROUPS

Treasure Hunt

Follow directions to find the hidden treasure!

Levels: Medium, Hard
Group size: 3–5

Before beginning

With adult help, put a string or rubber band on your right wrist.

Practice four directions to use in the treasure hunt:
  Take one step forward
  Take one step back
  Turn right (look at your wrist!)
  Turn left

Decide who will be the Finder. The Finder leaves the room while the others hide the treasure.

Materials

Per group
  small object to hide

Per child
  string or rubber band for each person's right wrist

Where should we put this toy cat so that Tomas doesn't see it right away when he comes back in the room?

1 Direct me!

The Finder returns to the room. The others take turns giving directions to the treasure.

Medium. Give one of the four directions you practiced.

Hard. Give one of the four directions you practiced. Change the number of steps (e.g., take three steps back).

Should Tomas go forward or back to get closer to the treasure?
Does it matter if he takes big steps or baby steps?
Which way is your right? Which way is Tomas's right?

2 Keep going until the Finder finds the treasure

Whoever gave the last direction gets to be the Finder next time.

Variations

Make your own directions (Hard). Add your own directions, for instance, turn to the window or turn half-way around.

Map It (Hard). Make a map to showing how to find the treasure.

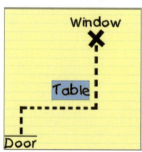

Anytime, Anywhere

Activities to do and games to play wherever you are: in the car, on the bus, in a waiting room, or at the dinner table.

Contents

Words on the Wall	92
Yes, No, Maybe	93
How Far Can You Go?	94
What's on the Page?	95
Seeing Shapes	96
Take Two	98

Anytime, Anywhere in other sections

Games

Hide, Share, Compare	13

Food and Water

Pretend Picnic	78

Good for Groups

Catch the Beat	82
Group Up	83
Stand and Vote	84
Who's Here?	88
Line Up	89

© 2014 TERC • Cambridge, MA

Words on the Wall

Levels: Medium, Hard

Group size: 2 (see Variations for more)

Waiting around? Play a game with words you see on the wall.

Before beginning

Pick a wall displaying a poster, sign, or bulletin board:

Medium. Choose something with a just few short words.

Hard. Choose something with a variety of words.

Decide who will be Player 1 first. Each player will have a turn.

1 Player 1 secretly picks a word

Player 1 gives a clue about the word.

2 Player 2 asks a yes-or-no question to try to identify the word

The question should be about number or position of letters. Do not ask if a certain word is the secret one.

Player 1 answers the question.

3 Keep asking and answering

What do you know so far? ... The third letter is E, and it has four letters. Which words on the poster fit the clues?

Player 2 may guess after asking two questions. If the guess is not correct, ask and answer again before another guess.

4 Switch roles and play again

If you were Player 1 last time, this time you're Player 2.

Variations

Play with a group (Hard). Play with three or four players. One player picks the secret word. The others take turns asking yes-or-no questions. The player who identifies the secret word wins.

Play on the page (Medium, Hard). Play with words on the page of a book or magazine.

Levels: Easy, Medium, Hard
Group size: any

Yes, No, Maybe

Will the next person to arrive have two eyes? … three eyes?
Predict what's possible.

Materials
none

1 Pose a silly or serious yes-or-no question

Ask about something that could happen …

Easy. … in the next few minutes.
  Will the next person to get on the bus have two heads?

Medium. … in the next hour.
  Will the rain stop by snack time?

Hard. … by the end of the day.
  Will we have a snowstorm today?

2 Predict

Everyone predicts "yes," "no," or "maybe."

Pick "maybe" only if you're really not sure!

 Why do you think we will have a snowstorm today? Do we usually get snow in the summer?

3 Wait and see

 Did your prediction come true?

Variations

Definite and impossible (Medium, Hard). Everyone says one thing that will (almost) definitely happen and one that is (almost) impossible. For instance, "When I wake up tomorrow morning, I will still be 5 years old," and "When I wake up tomorrow morning, I will have turned into a cat."

Number it (Hard). Choose a number from 1 to 10 to show the likelihood, with 1 unlikely or impossible and 10 very likely or definite.

© 2014 TERC • Cambridge, MA

ANYTIME, ANYWHERE 93

Levels: Easy, (Medium, Hard)
Group size: any

How Far Can You Go?

Can you cross the room in three kangaroo jumps? three monkey leaps?

1 Get in character

Choose an animal and practice moving like that animal. For instance, take bunny hops, duck waddles, or caterpillar crawls.

Materials

none

2 Predict

Stand to one side of the room. Then, predict how far you can go in three leaps, jumps, or other animal moves.

 Will you reach the chair with three frog leaps? Could you get all the way across the room?

I can cross the room in three frog leaps!

3 Go!

Make three moves.

4 Compare

 Did you go as far as you predicted? Do you think you can get there with one more waddle?

Variations

More moves (Medium, Hard). Predict how far you'll go in five (or ten) animal moves. Make your moves, and compare predictions and results.

Dress the part (Easy, Medium, Hard) Make a mask, ears, or a tail for your animal. Wear it when you see how far you can go.

Baby steps and giant steps (Easy, Medium). Start at one side of the room and take three baby steps. Then predict: will you go farther if you take three giant steps? Try it and see.

Baby step   Giant step

Levels: Easy, Medium, Hard

Group size: small enough so everyone has a chance to look and count

What's on the Page?

Count and discover ... in a book or magazine, or on a poster.

1 Find something to count on the image

Easy. Up to 4 similar items (e.g., birds, flowers).

Medium. Up to 6 similar items.

Hard. Up to 12 similar items.

Materials

Per group

a picture book, magazine, or almost any image around you

2 Count together

 I see flowers! How many are on the page? How can we keep track of which we've already counted?

3 Count again

 If we start on the bottom of the page instead of the top, do you think we'll get four again? Let's try it.

Variations

Count shapes (Easy, Medium, Hard). Count how many triangles (or circles, squares, or rectangles) are on the page.

Patterns on the Page (Easy, Medium, Hard). Find images or rhymes that show patterns. Predict what comes next in the pattern.

Count off the page (Easy, Medium, Hard). Count objects or shapes on a walk outdoors or at the playground.

How many in my picture? (Medium, Hard). Draw a picture with several apples, balloons, or trees. Trade pictures with a partner, who counts to find out how many you drew.

Compare counts (Hard). Count and compare two similar objects on the page.

Do you see more red balloons or more blue balloons?

ANYTIME, ANYWHERE

Levels: Easy, Medium, Hard

Seeing Shapes

Play a guessing game with shapes you see in the room.

Before beginning

Pick a section of the room:

Materials

none

Easy. Choose an area with 3-4 objects in different geometric shapes.

Medium. Choose an area with 5-6 objects in different geometric shapes.

Hard. Choose an area with a wide variety of objects and shapes.

Decide who will be Player 1 first. Each player will have a turn.

1 Player 1 secretly picks an object in the chosen area

2 Player 1 announces the shape

3 Player 2 asks a yes-or-no question to try to identify the object

Ask about features of the object. Do not ask if a certain object is the secret one.

Player 1 answers the question.

Group size: 2 (see Variations for more)

4 Keep asking and answering

 What do you know so far? … It's a circle, it's not blue and we use it to tell time. What could it be?

Player 2 may guess after asking two questions. If the guess is not correct, ask and answer again before another guess.

5 Switch roles and play again

If you were Player 1 last time, this time you're Player 2.

Variations

Play with winners (Hard). Keep track of how many questions each player asks to identify the secret object. The player who asks the fewest questions wins.

Play on the page (Easy, Medium, Hard). Play with images see as you look at a book or magazine.

Play with a group (Hard). Play with three or four players. One player picks the secret object. The others take turns asking yes-or-no questions. The player who identifies the secret object wins.

Play with blocks (Easy, Medium, Hard). Assemble a set of blocks of different shapes and colors. Ask yes-or-no questions to identify the secret block. After each question, remove the blocks that are ruled out.

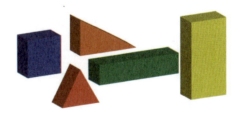

ANYTIME, ANYWHERE

Take Two

Time to clean up! Count as you put away toys on the floor.

Level: Easy (Medium, Hard)

Group size: 1 (see Variations for more)

1 Is two enough?

 If you put away two toys, will you clear the floor? Do you think there are more than two toys on the floor?

2 Take two

Count and put away two toys.

3 Are we done?

Keep "taking two" until everything is picked up.

4 How many in all?

An adult helps count by 2s to find the total.

Materials

Per group

ordinary clutter

    Easy. Up to 4 items.

    Medium. Up to 6 items.

    Hard. Up to 12 items.

Variations

Take five (or ten) (Medium, Hard). Try this with a group of 2–4. Take five (or ten) toys off the floor at a time. Then count by 5s (or 10s) to find the total.

Estimate first (Hard). Before counting, estimate the total amount on the floor. Then, count as you clean up.

Take shapes (Easy, Medium, Hard). Take all the toys that include circles off the floor. Then, take all the rectangles. Is anything left?

All Year Round

Ideas for mixing math into seasons, holidays, and special events all year round.

Contents

Spring. Grow, bloom, and blossom as
the earth comes alive. 100

Summer. Let your imagination
soar while you're cooling down. 101

Fall. Celebrate the start of the school year with
new friends, harvest, and Halloween. 102

Winter. Warm up with bright lights, holidays,
and a happy new year. 103

© 2014 TERC • Cambridge, MA

Spring

Grow, bloom, and blossom as the earth comes alive.

Games

Rabbit run
Play *Dinosaur Dash* (p. 10) with a toy rabbit for Rabbit Run, a cow for Cow Caper, or a bug for Bug Bustle.

Reuse, reduce, recycle
In honor of Earth Day, play *Same or Different* (p. 5) with recycled plastic bottle caps, old marker caps, or other materials you no longer need.

Projects and Crafts

Craft a card
Use shapes and words to design cards for Mother's Day, Father's Day, poetry month, or other spring holidays with *Picture Poems* (p. 50).

Get planting
Indoors or out, start with seedlings and track how they change from week to week with *Watch Me Grow* (p. 40).

Food and Water

Create a cookbook
Try the Smoothie cookbook variation of *The Counting Chef* (p. 80). Make copies of your cookbook to share with friends.

Mango smoothie for 1
1/2 cup diced mango
1/3 cup plain yogurt
1/4 cup orange juice
1 t agave syrup or honey
pinch of cinnamon

Smoothies for 2
1 banana
1 ¼ cups orange juice
½ cup frozen blueberries
5 frozen strawberries

Bag it
Celebrate fruits of the season by exploring their shapes, sizes, and textures with *In the Bag* (p. 78).

Good for Groups

Play outside
Got a group going outdoors? *Line Up* (p. 89) by height, hair length, or number of pockets.

Anytime, Anywhere

Act like an animal
Get down on all fours and get moving with *How Far Can You Go?* (p. 94).

100   ALL YEAR ROUND

© 2014 TERC • Cambridge, MA

Summer

Let your imagination soar while you're cooling down.

Games

Play with treasure
Collect shells, rocks, stickers, and coins on your summer adventures to use for a game of *Empty the Toy Box* (p. 14).

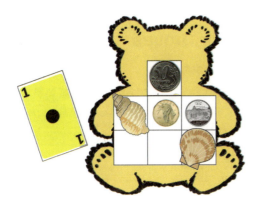

Spread the bounty
Make sure all the players get some treasure with *Share the Teddy Bears* (p. 4).

Projects and Crafts

Dream up a city
Build your city on the sidewalk. Use chalk to trace it with the **Try it outside** variation of *Cityscape* (p. 39).

Shrink it down to size
Create a mini castle, tower, or palace for toy animals to live in with *Towering Toothpicks* (p. 43).

Food and Water

Play at the pool
Cool off with *Fill It or Spill It* (p. 72) in a wading pool or big pool.

Pack it with sand
Bring some empty plastic containers to the sandbox or the beach for a sandy variation of *Water Wonders* (p. 74).

Good for Groups

Bury treasure in the sand
Hold a *Treasure Hunt* (p. 90) in the sandbox or at the beach.

Anytime, Anywhere

Look around you
Try *Seeing Shapes* (p. 96) when you're on the go: in the car, on the bus, or on a walk.

© 2014 TERC • Cambridge, MA

Fall

Celebrate the start of the school year with new friends, harvest, and Halloween.

Games

Learn about the group

Play the **Secret person** variation of *Secret Card* (p. 6) and get to know your group.

Stay busy as the nights get longer.

Play a different variation of *Match* (p. 2) every night of the week.

Projects and Crafts

Chart your growth

Find out how much you grow this school year. Make a strip of paper as tall as you are and then *Revise Your Size* (p. 34) at the end of the school year.

Make new friends

Make a friendship quilt with *Hands on the Wall* (p. 46).

Create a costume

Make a Mask (p. 41) and wear it for a costume party or for Halloween.

Food and Water

Have a happy harvest

Gather pumpkin, squash, and other harvest vegetables and fruits, and predict *What's Inside?* (p. 77).

Good for Groups

Compare costumes

What's your favorite costume? Does everyone have the same favorite? Find out with *Quick Questions* (p. 85).

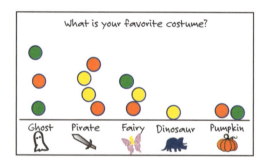

Anytime, Anywhere

Make a game of it

Starting the school year with new books? Explore shapes, sizes, and numbers with *What's on the Page?* (p. 95).

102   ALL YEAR ROUND   © 2014 TERC · Cambridge, MA

Winter

Warm up with bright lights, holidays, and a happy new year.

Games

Light up the night with stars in the sky
Play *Piggy Bank* (p. 8) with cutout stars or star stickers instead of pennies. For a valentine variation, play with hearts.

Celebrate 100 days or any day
Play *Dinosaur Dash* or the variation **Dinosaur dash 100** (p. 10) on the 100th day of school or the 100th day of the year.

Projects and Crafts

Wish for a happy new year
Make a *Picture Book* (p. 37) with a wishes theme. Show your own wishes or make a wish book for someone else.

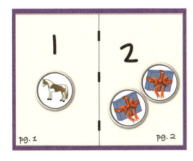

Decorate for the season
Use *Chain Challenges* (p. 42) to create paper chains for decorations. Create a color pattern in your chain and see if someone can discover it.

Food and Water

Help serve
At a party or family gathering, make sure everyone gets a fair share of a favorite treat with *One for You, One for Me* (p. 75).

Bring spring a little sooner
Let your imagination soar on a *Pretend Picnic* (p. 76). Take along your favorite fruits.

Good for Groups

Put on a show
Catch the Beat (p. 82) with a winter rhythm: jump over the ice, wave your fingers to sparkle like a snowflake, and call out brrrr! Then repeat: jump, wave, brrr … jump, wave, brrr.

Anytime, Anywhere

Foretell the future
Predict something that will happen in the coming year, something that won't happen, and something that might happen with *Yes, No, Maybe* (p. 93).

© 2014 TERC • Cambridge, MA

All Year Round

Math Connections

The games, projects, and activities in SAY IT WITH SHAPES AND NUMBERS span the key topics in the NAEYC/NCTM Joint Position Statement on Early Childhood Mathematics[1], and the Kindergarten Common Core State Standards for Mathematics[2].

Use the charts on pp. 106-109 to match the games, projects, and activities, along with their variations, to math content by level of difficulty:

- **Ages 3-4 (Easy)** reflects the key topics in the NAEYC/NCTM Joint Position Statement on Early Childhood Mathematics;

- **Ages 4-5 (Medium)** reflects the key topics in the NAEYC/NCTM Joint Position Statement on Early Childhood Mathematics;

- **Ages 5-6 (Hard)** reflects the key topics in the Kindergarten Common Core Standards for Mathematics.

Levels also reflect "typical" 3-4 year old, 4-5 year old, and 5-6 year old cognitive, social, language, fine motor, and other skills. Keep in mind that young children vary widely in abilities. Some people start with Easy for almost any age and then move up as needed.

SAY IT WITH SHAPES AND NUMBERS offers a rich foundation for engaging young children in exploring a wide range of mathematical topics, in appreciating the role of math in everyday life, and in establishing comfort and confidence in math—in school or out. You can use the activities, games, and projects in the book to prepare children for Pre-Kindergarten or Kindergarten, to complement any Pre-Kindergarten or Kindergarten math curriculum, and to integrate math into children's socializing, playing, and exploring. All children—no matter where they are along the developmental trajectory—will benefit.

---

[1] National Association for the Education of Young Children, & National Council of Teachers of Mathematics (2002, 2010). *Position statement. Early childhood mathematics: Promoting good beginnings.* Retrieved from http://www.naeyc.org/files/naeyc/file/positions/psmath.pdf.

[2] National Governors Association Center for Best Practices, & Council of Chief State School Officers. (2010). *Common Core State Standards for Mathematics.* Washington, DC: National Governors Association Center for Best Practices, & Council of Chief State School Officers.

Math Connections
(includes activity variations)

		Ages 3-4 (Easy)					Ages 4-5 (Medium)				
		Numbers and operations	Patterns and algebraic thinking	Measurement	Displaying and analyzing data	Geometry and spatial sense	Numbers and operations	Patterns and algebraic thinking	Measurement	Displaying and analyzing data	Geometry and spatial sense
Games	Match, p. 2	√			√		√			√	
	Share the Teddy Bears, p. 4	√	√				√	√			
	Same or Different?, p. 5	√					√				
	Secret Card, p. 6						√		√	√	√
	Piggy Bank, p. 8	√					√		√	√	√
	Dinosaur Dash, p. 10						√	√		√	√
	Hide, Share, Compare, p. 13	√					√		√	√	√
	Empty the Toy Box, p. 14						√	√	√	√	√
	Flip and Match, p. 16	√			√		√			√	
Projects and Crafts	Revise Your Size, p. 34			√					√		
	Strawberries and Blueberries, p. 35	√					√				
	Soaring Towers, p. 36	√		√		√	√		√		√
	Picture Book, p. 37	√	√				√	√			
	Fill It Up, p. 38	√		√		√	√		√		√
	Cityscape, p. 39		√	√	√	√		√	√	√	√
	Watch Me Grow, p. 40			√	√				√	√	
	Make a Mask, p. 41										√
	Chain Challenges, p. 42	√	√	√			√	√	√		
	Towering Toothpicks, p. 43			√		√			√		√
	Make It Morse, p. 44						√				√
	Hands on the Wall, p. 46	√	√	√		√	√	√	√		√
	Copy Cat, p. 47	√	√		√			√			√
	Say It with Shapes, p. 48		√		√			√	√		√
	Picture Poems, p. 50	√			√						√

Math Connections
(includes activity variations)

		Ages 5–6 (Hard)				
		Counting and Cardinality	Operations and Algebraic Thinking	Number/Operations in Base Ten	Measurement and Data	Geometry
Games	Match, p. 2	√			√	√
	Share the Teddy Bears, p. 4	√	√			
	Same or Different?, p. 5	√	√			
	Secret Card, p. 6	√			√	√
	Piggy Bank, p. 8	√	√	√		
	Dinosaur Dash, p. 10	√	√	√		
	Hide, Share, Compare, p. 13	√	√			
	Empty the Toy Box, p. 14	√	√			
	Flip and Match, p. 16	√	√		√	
Projects and Crafts	Revise Your Size, p. 34				√	
	Strawberries and Blueberries, p. 35	√	√			
	Soaring Towers, p. 36	√			√	√
	Picture Book, p. 37	√	√	√		
	Fill It Up, p. 38	√			√	√
	Cityscape, p. 39				√	√
	Watch Me Grow, p. 40				√	
	Make a Mask, p. 41					√
	Chain Challenges, p. 42	√		√	√	
	Towering Toothpicks, p. 43	√			√	√
	Make It Morse, p. 44	√	√			
	Hands on the Wall, p. 46	√	√		√	√
	Copy Cat, p. 47					√
	Say It with Shapes, p. 48					√
	Picture Poems, p. 50					√

© 2014 TERC • Cambridge, MA

Math Connections
(includes activity variations)

		Ages 3-4 (Easy)					Ages 4-5 (Medium)				
		Numbers and operations	Patterns and algebraic thinking	Measurement	Displaying and analyzing data	Geometry and spatial sense	Numbers and operations	Patterns and algebraic thinking	Measurement	Displaying and analyzing data	Geometry and spatial sense
Food and Water	Fill It or Spill It, p. 72	√		√		√	√		√		√
	Snack Station, p. 73	√					√				
	Water Wonders, p. 74			√		√			√		√
	One for You, One for Me, p. 75	√	√				√	√			
	Pretend Picnic, p. 76	√	√				√	√			
	What's Inside?, p. 77	√					√				
	In the Bag, p. 78						√				√
	My Height in Boxes, p. 79	√		√		√	√		√		√
	The Counting Chef, p. 80	√		√			√		√		
Good for Groups	Catch the Beat, p. 82		√					√			
	Group Up, p. 83	√		√	√		√		√	√	
	Stand and Vote, p. 84	√			√		√			√	
	Quick Questions, p. 85	√			√		√			√	
	Toy Store, p. 86	√					√				
	Who's Here?, p. 88	√			√					√	
	Line Up, p. 89	√		√	√		√				
	Treasure Hunt, p. 90			√		√			√		√
Anytime, Anywhere	Words on the Wall, p. 92						√			√	
	Yes, No, Maybe, p. 93		√		√			√		√	
	How Far Can You Go?, p. 94	√		√			√		√		
	What's on the Page?, p. 95	√			√	√	√			√	√
	Seeing Shapes, p. 96				√	√				√	√
	Take Two, p. 98	√	√			√	√	√			√

Math Connections
(includes activity variations)

		Ages 5-6 (Hard)				
		Counting and Cardinality	Operations and Algebraic Thinking	Number / Operations in Base Ten	Measurement and Data	Geometry
Food and Water	Fill It or Spill It, p. 72	√			√	√
	Snack Station, p. 73	√			√	
	Water Wonders, p. 74				√	√
	One for You, One for Me, p. 75	√	√			
	Pretend Picnic, p. 76	√				
	What's Inside?, p. 77	√			√	
	In the Bag, p. 78					√
	My Height in Boxes, p. 79	√			√	√
	The Counting Chef, p. 80	√	√		√	
God for Groups	Catch the Beat, p. 82	√				
	Group Up, p. 83	√			√	
	Stand and Vote, p. 84	√			√	
	Quick Questions, p. 85	√	√		√	
	Toy Store, p. 86	√	√	√		
	Who's Here?, p. 88	√	√	√	√	
	Line Up, p. 89	√			√	
	Treasure Hunt, p. 90				√	√
Anytime, Anywhere	Words on the Wall, p. 92	√			√	
	Yes, No, Maybe, p. 93				√	
	How Far Can You Go?, p. 94	√			√	
	What's on the Page?, p. 95	√			√	√
	Seeing Shapes, p. 96				√	√
	Take Two, p. 98	√	√			√

© 2014 TERC • Cambridge, MA

Letters

Letters

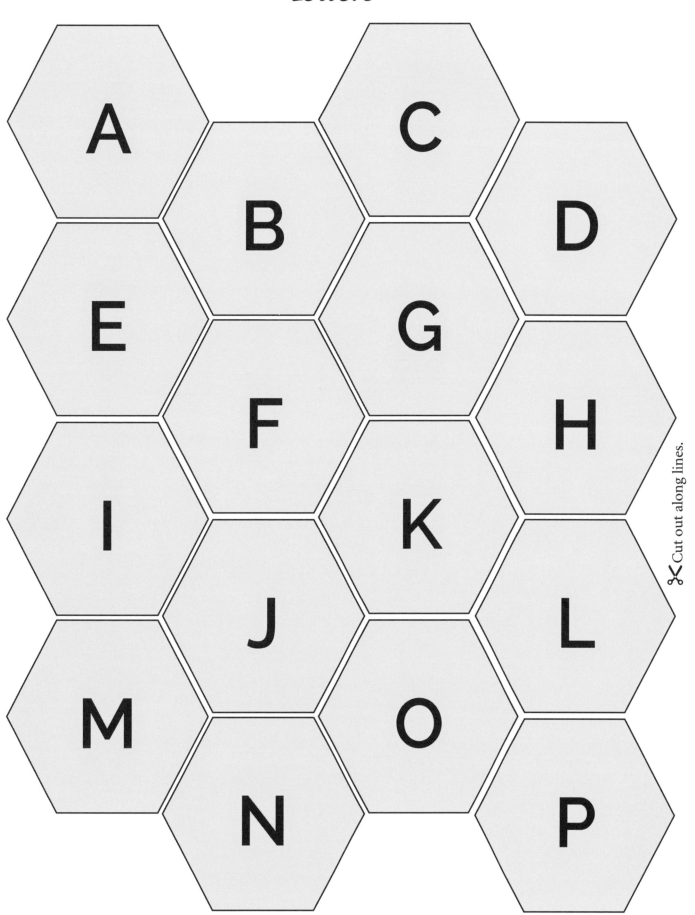

115

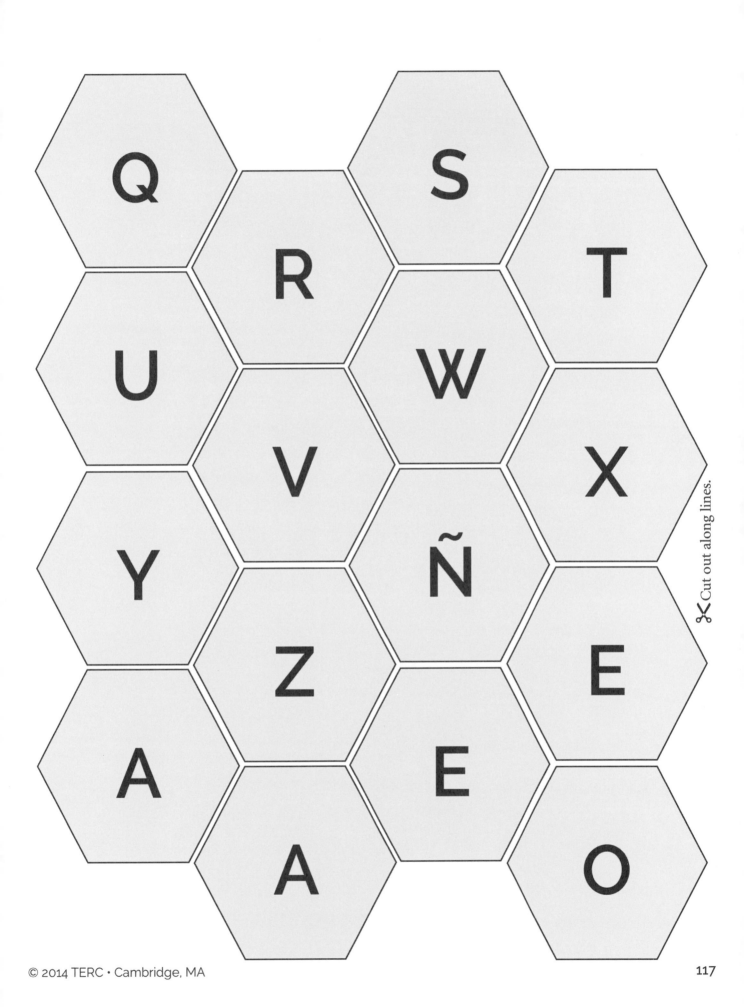

Squares—blank or fill in with nouns

✂ Cut out along lines.

© 2014 TERC • Cambridge, MA

Squares—blank or fill in with nouns

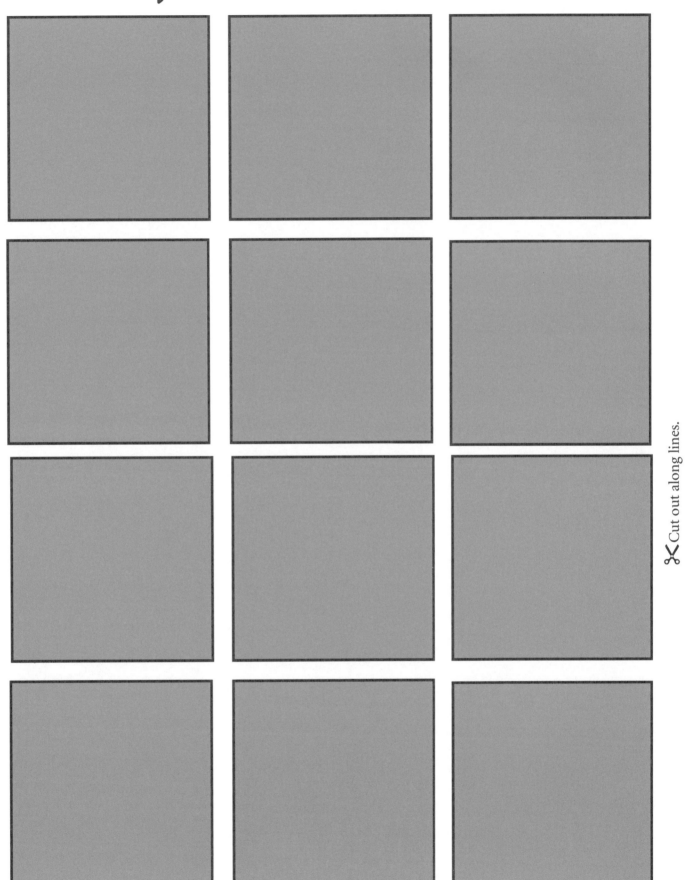

✂ Cut out along lines.

© 2014 TERC • Cambridge, MA

Rectangles—blank or fill in with exclamations

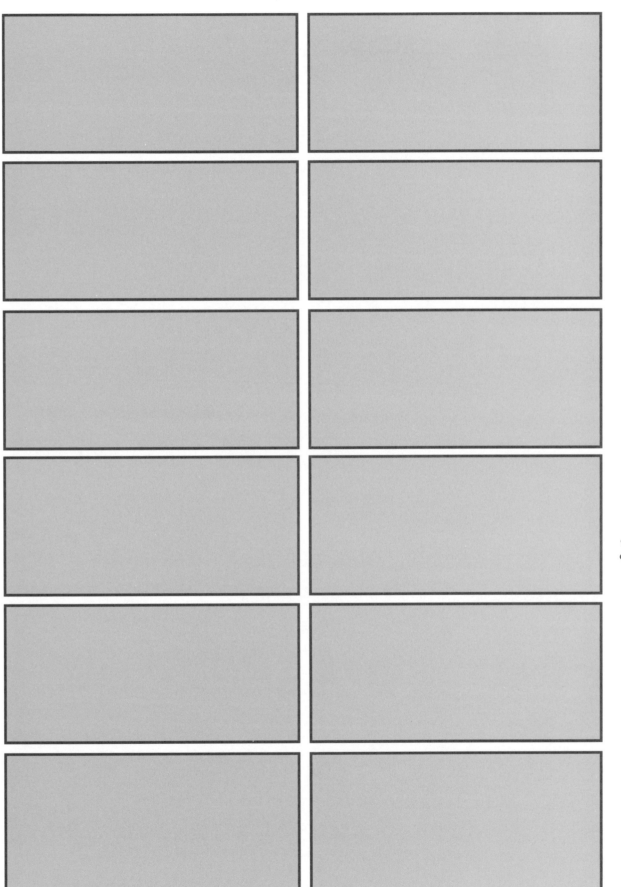

Cut out along lines.

© 2014 TERC • Cambridge, MA

123

Rectangles—blank or fill in with exclamations

Cut out along lines.

© 2014 TERC • Cambridge, MA

125

Triangles—blank or fill in with adjectives

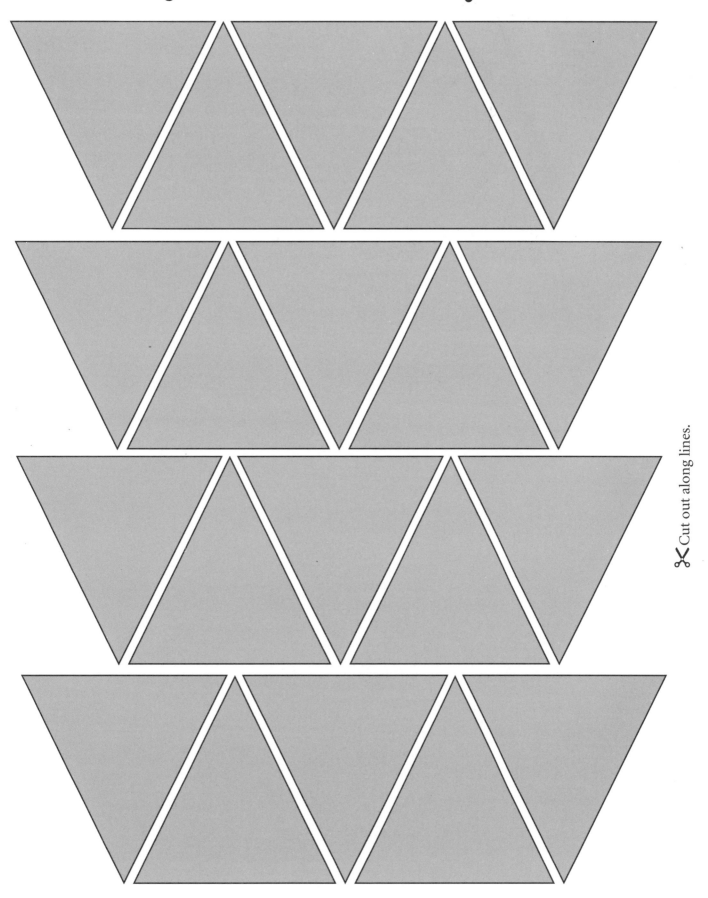

Triangles—blank or fill in with adjectives

Cut out along lines.

Hexagons—blank or fill in with letters

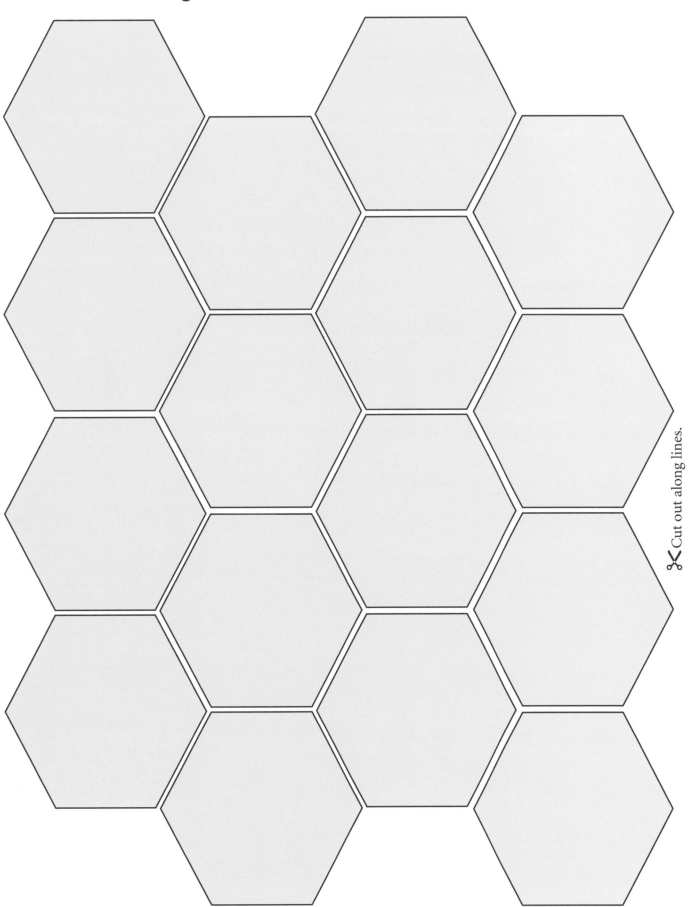

Hexagons—blank or fill in with letters

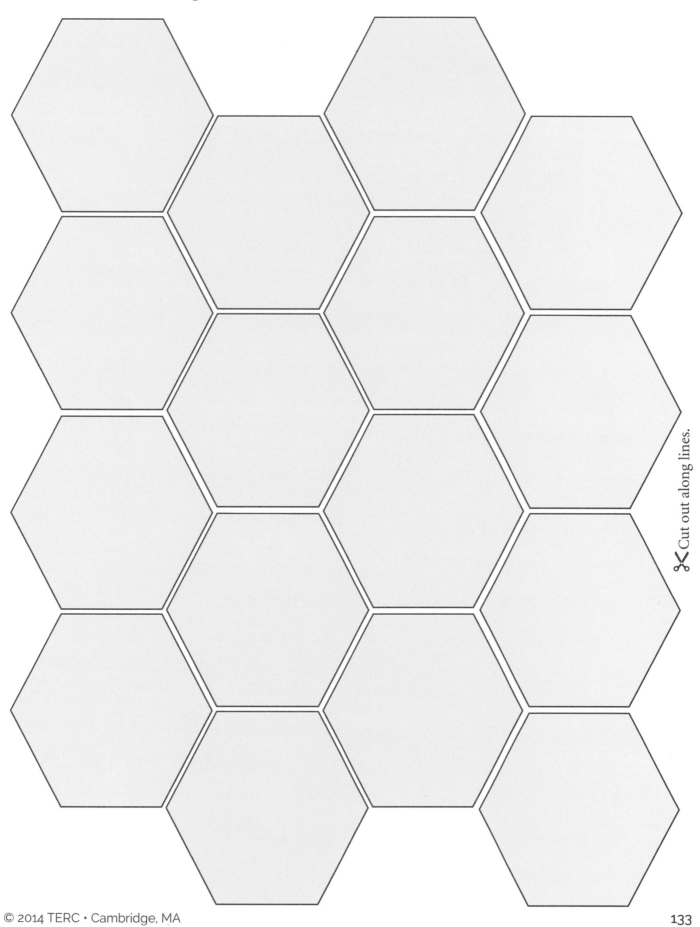

Cut out along lines.

© 2014 TERC • Cambridge, MA